DETECTION OF
AND CONSTRUCTION AT
THE SOIL/ROCK INTERFACE

Proceedings of the Symposium sponsored by
the Rock Mechanics Committee of
the Geotechnical Engineering Division of
the American Society of Civil Engineers
in conjunction with the ASCE
Convention in Orlando, Florida

October 24, 1991

Edited by William F. Kane and Bernard Amadei

Geotechnical Special Publication No. 28

Published by the
American Society of Civil Engineers
345 East 47th Street
New York, New York 10017–2398

ABSTRACT

This proceedings, Detection of and Construction at the Soil/Rock Interface, consists of papers presented at the Symposium held on October 24, 1991 in conjunction with the ASCE National Convention, Orlando, Florida. It deals with a discussion of where rock starts, and the engineering and construction problems that occur at this boundary. This topic is especially important to consultants and contractors. Papers are of a practical nature covering a wide perspective. Topics include a keynote discussion of the soil/rock boundary followed by discussions and case studies on the definition of rock as it pertains to construction and highway projects. Subjects covered include geographic differences in the definition of rock, effects of equipment changes on the definition, and ground stabilization in soft ground tunneling.

Library of Congress Cataloging-in-Publication Data

Detection of and construction at the soil/rock interface: proceedings of the symposium/sponsored by the Rock Mechanics Committee of the Geotechnical Engineering Division of the American Society of Civil Engineers in conjunction with the ASCE Convention in Orlando, Florida, October 24, 1991; edited by William F. Kane and Bernard Amadei.
     p.   cm. — (Geotechnical special publication; no. 28)
Includes indexes.
ISBN 0-87262-849-3
    1. Engineering geology — Congresses. 2. Engineering geology — Case studies — Congresses. I. Kane, William F. II. Amadei, Bernard, 1954-   . III. American Society of Civil Engineers. Geotechnical Engineering Division. Rock Mechanics Committee. IV. ASCE National Convention (1991: Orlando, Fla.) V. Series.
TA703.5.D48   1991
624.1'51—dc20   91-30343
                               CIP

The Society is not responsible for any statements made or opinions expressed in its publications.

PREFACE

The problem of defining just what rock actually is, and then building on this material, has increasing importance in civil engineering construction. Economic considerations play the major role. Costs are based on what a contractor is excavating: is it soil or rock? Often this is a matter of equipment. One contractor with a D10 dozer and ripper treats the material as soil, while his under-powered competitor must blast.

The definition of soil/rock also is a problem as conditions change. Litigation due to changing conditions is now commonplace. In fact, the Association of Drilled Shaft Contractors (ADSC) is so concerned with this issue that it sponsors numerous roundtable discussions and seminars on the topic.

This publication contains a collection of case studies describing the problems associated with defining, and constructing on, the soil/rock boundary. It begins with an attempt to develop a working definition of rock, follows this with numerous case studies, and then looks at attempts to define it contractually and in the laboratory.

This symposium was sponsored by the Rock Mechanics Committee, Geotechnical Division of the ASCE. It is the current practice of the Geotechnical Engineering Division that each paper published in a special publication be reviewed for its content and quality. These special publications are intended to reinforce the programs presented at convention sessions or specialty conferences, and to contain papers that are timely or controversial to some extent. Ordinarily the reviews are carried out within a three-month period. The standards of review are essentially those for the ASCE Journal of Geotechnical Engineering, but the exigencies of timeliness and the need to have the publication available at the convention preclude more than one cycle of editing and revision. Therefore, it should be recognized that there are some differences in purpose between contributions to the special publications and those in the Journal. All papers are eligible for ASCE awards as well as formal discussion in the Journal of Geotechnical Engineering.

The editors would like to thank the Rock Mechanics Committee for its suggestions, the paper authors for their cooperation and effort under a tight schedule, and the reviewers for taking the time to do thorough, constructive reviews. Reviewers for this publication were:

Richard M. Bennett   Theodore L. Triplett
B. Alex Grenoble   John P. Turner
Robert D. Hatcher, Jr.   Bernard Amadei
Gregg A. Scott   William F. Kane
Barry Thacker

Cover photos were provided by Charles H. Dowding and William F. Kane.

William F. Kane   Bernard Amadei
Department of Civil Engineering   Department of Civil and Environmental Engineering
University of the Pacific   Colorado University

CONTENTS

GEOTECHNICAL SPECIAL PUBLICATIONS

THE SOIL-ROCK BOUNDARY: WHAT IS IT AND WHERE IS IT?

Fred H. Kulhawy[1], F.ASCE, Charles H. Trautmann[2], M.ASCE, and Thomas D. O'Rourke[1], M.ASCE

ABSTRACT: A major issue in many areas of Civil Engineering practice is the boundary between earth materials that are defined as soil and those that are defined as rock. Designers, contractors, and attorneys are particularly sensitive to this issue. This paper attempts to put major aspects of this problem in perspective. A general framework is presented for describing earth materials in a consistent manner, and examples are given to illustrate the nature of the problem involved.

INTRODUCTION

Horror stories abound in the construction industry over rock that wasn't where it was supposed to be or that appeared unexpectedly. Unfortunately, these situations precipitate adversary positions between the designer and contractor that, all too often, are resolved in a court of law. Most of the time, these problems could have been resolved if there had been open communication among all concerned parties. Fundamental to all communication is language, including choice and meaning of words. Soil and rock are two terms that are used rather casually, and most people think they know what these materials are. But is this belief really true? And is one person's understanding the same as the next? And does rock mean the same thing to a designer in New York, whose project (and rock) is in Florida, and a contractor from Illinois who is bidding the project? Much of the time, the answer to these questions is no, unfortunately.

In this paper, some of the key issues involved in these misunderstandings over "soil", "rock", and the "soil-rock boundary" are examined. Then a simple, consistent, and unambiguous approach is proposed that should eliminate these misunderstandings. Examples are given where appropriate to illustrate some aspects of the problem involved.

---

1 - Professor, School of Civil and Environmental Engineering, Hollister Hall, Cornell University, Ithaca, NY 14853-3501
2 - Senior Research Associate (at Cornell affiliation above) and Executive Director, The Sciencenter, Ithaca, NY 14850

SOME BASIC DEFINITIONS

What is soil? And what is rock? To begin to answer these questions, some pertinent definitions were extracted from five well-known reference sources and are summarized in Table 1. Soil and rock were defined in all five sources, but bedrock was defined in only four of them. In the Dictionary of Geological Terms (AGI, 1962), hard rock and soft rock also were defined explicitly.

Consider the soil definitions. While there are obvious differences in detail, overall there is a striking similarity. In all cases, soil is defined as the mass that is composed of some variety of material types. Based on this consistency, an overall working definition would be as follows:

> Soil The surficial layer or mantle of fragmentary earth material, of whatever origin, that nearly everywhere forms the surface of the land and rests on the bedrock. It includes residual, transported, organic, and other natural materials.

One may choose to define a topsoil as well, indicating that the top tens to hundreds of millimeters of the soil are rich in organic material. Usually this differentiation is of minor concern for Civil Engineering purposes.

However, when viewing the definitions for rock in Table 1, it becomes apparent that there are stark differences among the reference sources. The definitions range from the rock material to the in-situ rock mass and from very sound or consolidated to unconsolidated rock material or mass. In addition, the Dictionary of Geological Terms (AGI, 1962) defines the rock and its hardness as a function of excavation technology. With the inconsistencies cited above and in the table, there is no question that misunderstandings would develop over what is called "rock".

Figure 1 provides a reference frame for evaluating the generic rock terminology. First, one must consider the unweathered rock material, which can be defined simply as "naturally formed, coherent

Table 1. Selected Definitions for Soil and Rock

WEBSTER'S UNABRIDGED DICTIONARY (Gove, 1966)

Soil (a) Upper layer of earth that may be dug or plowed. (b) Surface earth of a particular place with reference especially to its composition or its adaptability (as for the farmer, builder, or engineer).

Rock Consolidated or unconsolidated solid mineral matter composed of one or usually two or more minerals or partly of organic origin (as coal) that occurs naturally in large quantities or forms a considerable part of the earth's crust.

Bedrock Solid rock underlying the soil and other unconsolidated materials or appearing at the surface where these are absent.

(continued)

Table 1. Selected Definitions for Soil and Rock (completed)

RANDOM HOUSE UNABRIDGED DICTIONARY (Flexner, 1987)
Soil Portion of the earth's surface consisting of disintegrated rock and humus.
Rock Mineral matter of variable composition, consolidated or unconsolidated, assembled in masses or considerable quantities in nature, as by the action of heat or water.
Bedrock Unbroken solid rock, overlaid in most places by soil or rock fragments.

DICTIONARY OF GEOLOGICAL TERMS (AGI, 1962)
Soil The term soil is equivalent to regolith. (engineering geology)
Regolith Mantle rock; saprolith. The layer or mantle of loose, incoherent rock material, of whatever origin, that nearly everywhere forms the surface of the land and rests on the hard or "bed" rocks. It comprises rock waste of all sorts, volcanic ash, glacial drift, alluvium, windblown deposits, vegetal accumulations, and soils.
Rock (a) Strictly, any naturally formed aggregate or mass of mineral matter, whether or not coherent, constituting an essential and appreciable part of the earth's crust. (b) Ordinarily, any consolidated or coherent and relatively hard, naturally formed mass of mineral matter; stone. (c) To the engineer, the term rock signifies firm and coherent or consolidated substances that cannot normally be excavated by manual methods alone.
Hard Rock Rock which requires drilling and blasting for its economical removal. Loosely used to distinguish igneous and metamorphic from sedimentary rock.
Soft Rock Rock that can be removed by air-operated hammers, but cannot be handled economically by pick. Loosely used to distinguish sedimentary from igneous and metamorphic rock.
Bedrock Any solid rock exposed at the surface of the earth or overlain by unconsolidated material.

PENGUIN DICTIONARY OF CIVIL ENGINEERING (Scott, 1980)
Soil Soil is gravels, sands, silts, clays, peats and all other loose materials including topsoil, down to bedrock.
Rock A mass of different grains cemented together by a matrix.
Bedrock A loose term meaning hard rock underlying gravel or other loose surface soil.

CANADIAN FOUNDATION ENGINEERING MANUAL (CGS, 1985)
Soil That portion of the earth's crust that is fragmentary, such that some individual particles of a dried sample can be readily separated by agitation in water; it includes boulders, cobbles, gravel, sand, silt, clay, and organic matter.
Rock Natural aggregate of minerals that cannot be readily broken by hand and will not disintegrate on its first drying and wetting cycle.

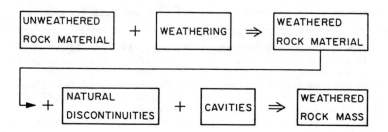

Figure 1.   General Progression of "Rock" Terminology

aggregate of mineral matter".   This material is not "solid", as is
often thought or spoken of in the popular press.   It contains a natural void structure that could range from micro to macroscopic and
that could be either continuous or discontinuous.   The material also
contains inherent fissures that can range from visible to minute.
Goodman (1976) stresses that these fissures can dominate the so-
called rock material strength and deformation properties.   More will
be said about this matter later.

Once the unweathered rock material has been defined, the in-situ
weathering profile must be evaluated to assess the characteristics of
the weathered rock material.   It is, of course, quite important to
differentiate between unweathered and weathered rock materials.   Procedures for assessing the degree of weathering are well-established
in the literature (e.g., Goodman, 1976; Legget and Hatheway, 1988).

After establishing the rock material conditions, the rock mass can
be defined.   As shown in Figure 1, the natural discontinuities and
cavities are included with the rock material to yield the weathered
rock mass.   The discontinuities can be defined as geologic features
that separate blocks of rock material.   Commonly, they include
joints, faults, bedding planes, cleavage planes, shear zones, etc.
These features can influence the overall engineering behavior dramatically, so that the rock mass commonly has a lower strength and
higher compressibility and permeability than the rock material.   The
strength change could be by a factor of two to four, the compressibility change could be an order of magnitude, and the permeability
change could be several orders of magnitude.   In addition, one must
be cautious of two specific types of discontinuities that have developed a rather notorious reputation of leading to engineering problems.   These two are foliation shears in metamorphic rocks and bedding plane shears in sedimentary rocks.   Potential problems with
these shear zones are described well in the literature (e.g., Deere,
1975; Deere and Varde, 1986).

In addition to the discontinuities, cavities may be present in
certain types of rock materials.   Solution cavities are most common
in limestones, but they also can be present in dolostone, marble,
gypsum, anhydrite, and salt.   Another type of cavity is the lava tube
found in volcanic flow rocks.   In any of these cases, the presence of
the cavities can greatly alter the engineering properties of the rock
mass.   These features must be identified during exploration; other-

wise, unwelcome surprises can occur. One extreme example of this
problem was given by Deere and Varde (1986), as quoted below:

"In a recent case, a boring was being made at a limestone
site to check on possible karstic openings. None was found;
however, the core was stored in a nearby core storage shed.
When the pile of boxes reached a certain height, they did a
fine bit of exploration. They suddenly disappeared from
sight, having broken through the roof of a large, unknown
cave!"

Returning now to Figure 1, it should be quite clear that the term
"rock" should not be used unmodified for any technical purposes.
Any statement should be specific whether the rock material or rock
mass is being considered and whether it is weathered or unweathered.

IN-SITU PROFILE

With the above clarification of terminology, one can now look at
a general in-situ profile of earth materials, as shown in Figure 2.
The left column depicts the residual profile conditions in a region
that has not been glaciated. The next four columns depict the stratigraphic differentiations of three authors. Although there is a
difference in notation, the implications are consistent. With increasing depth, the profile changes from soil at the surface, to soil
with weathered rock fragments, to the weathered rock mass, and final-
ly to the unweathered rock mass. As noted in the sixth column, the
two soil mass units will have engineering properties and resulting

Idealized Profile	Sowers(1963) Igneous & Metamorphic	Deere & Patton (1971) All Rocks		Dearman(1976) All Rocks	Engineering Properties & Behavior	General Profile
		I	IA	VI		Topsoil
	Soil		A-Horizons	Soil or True Residual Soil	Soil Structure Controlled	
		Residual Soil	IB B-Horizons			Soil
			IC	V		
	Saprolite		C-Horizons (Saprolite)	Completely Weathered		
		II	IIA Saprolite to Weathered Rock Transition	IV Highly Weathered	Relict Discontinuity Controlled	
	Partially Weathered Rock	Weathered Rock				Weathered to Unweathered Rock Mass (Bedrock)
			IIB Partly Weathered Rock	III Moderately Weathered		
		III		II Slightly Weathered	Discontinuity Controlled	
	Solid Rock	Unweathered Rock				
				I Fresh Rock		

Figure 2. Some Comparisons of Weathering Profiles
(Based on summary by Irfan and Cipullo, 1990)

behavior under load that is controlled by the soil structure. The
various soil structures are described by Kulhawy, et al. (1989). The
properties and behavior of the two rock mass units normally are
controlled by the discontinuities or, perhaps, cavities. These features were described previously. In the upper, more weathered zones,
relict features still control the properties and behavior.

In glaciated regions, the residual soils and weathered portions of
the rock mass commonly have been stripped away. In these cases, the
profile usually is simpler, with a surficial transported soil underlain by a more-or-less unweathered rock mass. The extent of weathering usually is small.

For both of the above conditions, the profile can be generalized
as shown in the right column of Figure 2. There is possibly some
topsoil, underlain by a soil mass that has its engineering properties
and behavior controlled by the soil structure. Below the soil is the
weathered to unweathered in-situ or native rock mass that has its
engineering properties and behavior normally controlled by the discontinuities or cavities. This rock mass is the bedrock.

Returning again to Table 1, it can be seen that bedrock is described as solid or unbroken or hard. In fact, as noted by Legget
and Hatheway (1988), "bedrock itself seems to connote a condition of
sturdiness and incompressibility". However, as described in conjunction with Figure 2, bedrock is actually the in-situ or native rock
mass below the soil. It may be weathered or not, and its characteristics will depend upon the rock material, discontinuities, and
cavities, if present.

UNWEATHERED ROCK MATERIAL

Many of the classifications for either the rock material or the
rock mass include the uniaxial compressive strength of the unweathered rock material as an integral component. Figure 3 presents a
summary of some of the better-known classifications used to describe
the unweathered rock material strength. The boundaries and terminology vary considerably, so there is undoubtedly some possibility of
misunderstanding occurring just because of these differences.

At the high end of the strength scale, the rock material is
extremely strong, and there is little problem likely to be encountered with this material in Civil Engineering practice, except that
it may be costly to excavate. At the lower end of the strength
scale, the situation is dramatically different. The rock material is
weak and, when weathering, discontinuities, or cavities are superim-
posed, the rock mass is even weaker. Either the rock material or the
rock mass is likely to control the overall geotechnical design.

To establish a convenient reference boundary, it is worthwhile to
consider the strength of the rock material with respect to that of
concrete. The typical lower strength of concrete used in construction is 20 MN/m^2 (3 ksi), while the upper strength of concrete used
is on the order of 100 MN/m^2 (15 ksi). These values provide convenient boundaries between strong, medium, and weak rock materials, as
denoted by "proposed" in Figure 3.

The strong rock material always will be stronger than concrete,
and it is not likely that even the overall rock mass would be weaker

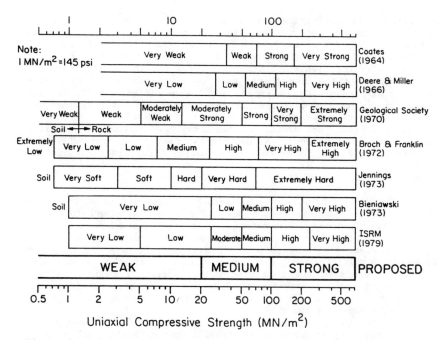

Figure 3.   Classifications for Unweathered Rock Material Strength
            (Summary by Bieniawski, 1980)

than concrete.   For the intermediate rock material, much depends on
the weathering, discontinuities, and cavities.   The resulting rock
mass could be either weaker or stronger than concrete.   With weak
rock material, the design will always be controlled by either the
rock material or mass features.   It should be noted that no lower
bound is proposed for weak rock because some porous rock materials
can be weaker than some heavily overconsolidated soil materials.

Much attention has been focused on both weak rock materials and
masses over the last fifteen years (e.g., Deere, 1975; Meigh and
Wolski, 1979; Akai, et al., 1981; Deere and Varde, 1986; Hatheway,
1990; Leroueil and Vaughan, 1990).   It is well-understood that these
materials can be very troublesome, and they need to be addressed
carefully.   Hatheway (1990) has even suggested the following definition for weak rock (mass):

"a consolidated earth material possessing an unusual degree
of bedding or foliation separation, fissility, fracturing,
weathering, and/or alteration products, and a significant
content of clay minerals, altogether having the appearance of
a rock, yet behaving partially as a soil, and often
exhibiting a potential to swell or slake, with the addition
of water; some weak rocks are also subject to time-dependent
release of stored tectonically-induced stress"

Although this definition may be all-inclusive and precise, it is perhaps a bit unwieldy. The simple strength definition given in Figure 3 (< 20 MN/m^2 or 3 ksi) is probably adequate. If the rock material is weaker than concrete, it is potentially troublesome.

For reference, Table 2 lists the general rock types that commonly would fit into the unweathered weak rock material category. With appropriate induration, cementing, etc., some of these rock materials could move into the medium or strong categories. With weathering, many other rock materials could become weak.

Table 3 lists some useful general indicators of whether weak rock materials may be present. Although these factors are not sufficient to define weak rock, they do provide useful first-order guidance.

ROCK MASS BEHAVIOR IN ENGINEERED CONSTRUCTION

As described in the previous pages, the general in-situ profile of earth materials consists of a soil layer that is underlain by the rock mass or bedrock. The engineering properties and behavior of the soil fundamentally are controlled by the soil structure, while those of the weathered or unweathered rock mass normally are controlled by the discontinuities or, perhaps, cavities. In the upper, more weathered zones of the rock mass (where present), relict features still control the properties and behavior. Furthermore, in weaker rocks, the material itself can contribute substantially to the overall behavior of the mass.

Once the in-situ profile has been defined, the issue that is perhaps more important presents itself: how will the soil or rock mass behave within a specific construction situation? In cases where there is a clear demarcation between the soil and rock masses in terms of material type and engineering properties, there is little misunderstanding that occurs among designers, contractors, and attorneys. However, when the demarcation is not clear, then problems can arise.

Table 2. General Types of Unweathered Weak Rock Material

Main Group	Rock Material
sedimentary	claystones and compaction shales friable sandstones chalks, marls, and calcarenites evaporites coal
igneous (porous volcanic or volcaniclastic)	tuff, agglomerate, and flow breccia
metamorphic	schists and phyllites containing weak platey minerals (chlorite, talc, etc.)

Based on Deere (1975) and Deere and Varde (1986)

Table 3. Some Indicators of Possible Unweathered Weak Rock Material

Cretaceous or younger age rocks
Permian or younger carbonate rocks
organic depositional origin
lack of cementing agents
silt- or clay-rich
greenish colors (e.g., chlorite)

Based on Hatheway (1990)

Figure 4 presents a useful overview of the issues to be addressed
in considering overall rock mass behavior. Focus is placed on the
rock mass because it is believed that the "rock" is the more
misunderstood quantity at the "soil-rock boundary".

The first issue is scale, and the question is whether the geologic
feature (discontinuity, cavity, etc.) is significant with regard to
the size of the constructed entity. If so, it is influential and
must be addressed explicitly. For example, discontinuities rarely
have negative impact on the stability of exploration boreholes (on
the order of 50 mm or 2 inch diameter) at typical engineering depths.
However, with a five meter (16 ft) diameter hole, these same discontinuities could cause major instability problems. With soil, scale
is almost always significant.

Second is the depth effect. At shallow depths, rock mass failure
is largely discontinuity-dominated, with rock block movement, discontinuity slippage, etc. At significant depths, rock mass failure is
largely stress-dominated, leading to plastic yield, spalling, etc.
Parallel types of behavior occur in soil masses.

The next three issues - stress state, ground water, and construction practices - are interrelated to some degree. In each case,
control is the key word. If the stress influences can be controlled,
if the ground water flow can be controlled, and if construction move-
ments or "damage" can be controlled, then the soil or rock mass can
perform well.

All of the above considerations need to be evaluated in the
context of a specific engineering or construction application. For
example, consider the case of excavation. With the lump sum, unit
price type of contract that tends to be favored in U.S. practice, the

Figure 4. Rock Mass Behavior Issues
(Based on Hudson, 1989)

definitions of soil and rock basically become economic distinctions
that determine pay quantities. For projects including large surface
excavation volumes, it is common to find specifications that define
soil and rock excavation on the basis of excavation equipment and
procedures needed for each (e.g., Goodman, 1976). While these
criteria may be expedient and practical, they are applicable only for
the purpose directly intended.   They do not define the soil-rock
boundary. The actual boundary being defined normally will be within
the weathered rock mass (or perhaps within an unweathered weak rock
mass). Also implicit in these definitions is that the scale of the
geologic features is significant, shallow depth effects dominate,
and stress state, ground water, and construction practices are of
little influence.
    When this excavation issue is considered in the context of drilled
shaft foundations, specification definitions of rock based on excavation technology are less appropriate. For example, a recent round
table meeting (Litke, 1991) of foundation drilling contractors and
geotechnical engineers concluded that "tieing specifications to a
brand or type of equipment or penetration rate is not a practical
method for determining a definition of rock".
    These excavation-based definitions also have been used to differ-
entiate between hard and soft rock, as given in Table 1 from the
Dictionary of Geological Terms (AGI, 1962). These types of definitions are of little specific use because they do not address the
issues in Figure 4.
    When this excavation issue is considered in tunneling, the situa-
tion changes again. The soil-rock profile definitions inherently are
based on a vertical reference frame, as are drillability, hardness,
etc. during exploration. However, tunneling represents a more-or-
less horizontal construction application. Simple excavation-based
definitions could even be misleading in this context because the
excavation procedure is controlled largely by the strongest mass
(horizontally) at the face, while the support system is controlled
largely by the weakest mass (vertically) in the crown. What is most
appropriate in this case is to describe the soil and rock masses as
given previously in this paper, evaluate their behavior in the
context of the issues in Figure 4, and then assess their probable
performance for the particular application. One recent example of
this general approach to evaluate the soil and weathered rock masses
in a realistic manner is given by Desai and Ku (1981) for the
Baltimore Metro. This type of approach is to be preferred in most
applications.

SOME ILLUSTRATIVE EXAMPLES

    To bring out some of the points made already, it is useful to
consider several pertinent examples. These examples typically would
be in the weak rock mass category.
    Young Limerock. In South Florida and other geologically similar
areas, young limerock/soil sequences exist that represent composite
soil-rock profiles. Figure 5 shows an illustrative profile typical
of south and east Florida. As shown, there are upper, middle, and
lower limerock sequences, interlayered with cemented sands. These

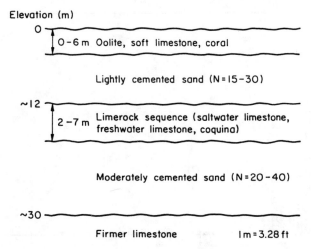

Figure 5. A South Florida Soil-Rock Profile

sequences, as shown, do not fit the classical patterns shown previously.

The lower limestone generally represents the top of "true" bedrock, while the upper and middle limerocks are rock layers within
the soil. These layers are rather complex in their origin, variable
in thickness, and variable in rock type and properties. On some projects, as many as six distinct horizons have been found within the
middle layer. Typically both of these layers are vuggy, and the vugs
are sand-filled. On one project, more than 300 uniaxial compression
tests were conducted on the middle layer rocks, giving the histogram
shown in Figure 6. The mean of these data is on the order of 5 MN/m^2
($\approx$ 750 psi). However, the extreme variability is quite evident. The
lower end of the data even might suggest soil, but the high end is
clearly a medium rock material.

To deal with this profile properly, one must appreciate the
profile and its variability. Then the behavior issues and specific
application can be addressed readily. Sometimes soil issues dominate
while, other times, rock issues dominate.

Volcaniclastics. Volcaniclastic deposits represent a class of
materials with both soil and rock characteristics. Their unusual
properties are frequently related to a porous assemblage of particles
that are partially cemented with volcanic glass and its weathered
derivatives.

An interesting example of this type of material has been described
by O'Rourke and Crespo (1988). They studied a volcaniclastic formation, known as Cangahua, which is found in the Andes of Ecuador and
Columbia. The uniaxial compressive strength of this material ranges
from 400 to 700 kN/m^2 ($\approx$ 60 to 100 psi). Using Figure 3, this material might be classified as soil; however, it possesses distinctive
rock-like characteristics, such as a relatively high tensile
strength. This tensile strength can sustain steep natural and exca-

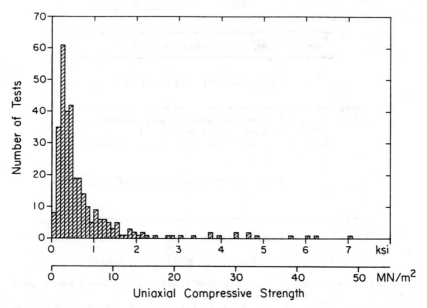

Figure 6.   Uniaxial Compressive Strength of Soft Florida Limerock

vated slopes of 60 to 80°, as high as 40 to 50 m (130 to 165 ft).
Moreover, the slope failure mechanisms do not conform to the shear
failure models of soil slope stability analysis.   Instead, they are
controlled by fracture nucleation and propagation from zones of high
tensile stress.

An interesting feature of many materials that border between rock
and soil is the relatively delicate and vulnerable nature of the
rock-like components of strength.   Figure 7 shows the uniaxial
compressive and Brazil tensile strengths of Cangahua as a function of
degree of saturation.   There is a clear relationship shown, with the
tensile strength declining to 25 percent of maximum as the degree of
saturation increases from 40 to 90 percent.   Such dramatic changes
can have a profound influence on the stability of slopes, and they
provide ample illustration of the sensitive nature of the in-situ
strength.

Shale.   It has long been recognized that shale can be a troublesome material (e.g., Goodman, 1976; Legget and Hatheway, 1988).
Although a detailed discussion of shale is well beyond the scope of
this paper, it is instructive to illustrate the behavior of some
shales by drawing attention to one in particular.

Substantial amounts of tunneling and open cut construction for
the Superconducting Super Collider (SSC) are planned in the Eagle
Ford Shale, a Cretaceous marine shale that outcrops in the Dallas-
Fort Worth area of Texas.   The SSC is a high-energy particle
accelerator that will be the largest scientific machine ever built,
involving 87 km (54 mi) of tunnel.   Recent studies (Earth Technology

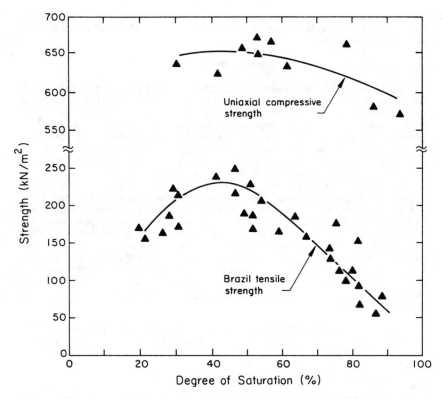

Figure 7.   Strength-Saturation Behavior of Cangahua Volcaniclastics

Corp., 1990) show an average uniaxial compressive strength of this shale equal to 1.95 MN/m^2 (283 psi), with a coefficient of variation = 0.51 for a sample of 123 tests.   In Figure 3, this material could be intermediate between soil and rock by some authors.

Tunneling at depths of 50 to 60 m (165 to 200 ft) involves stresses comparable to the shale compressive strength.   Accordingly, tunnel boring machines must be designed for continuous support, with full segmental linings erected as machines advance.   Open cut construction for the experimental halls, some of which may be 70 m (230 ft) deep, pose special problems.   Because of the potential for swelling and weakening, the base of the excavation needs to be protected from water and covered.   To prevent buckling of the floor and reduce heave, large diameter anchors or shafts may be advisable to tie down the excavation floor.   Dry drilling will reduce exposure to water.

For a project with the size and sensitivity of the SSC, characterization of the shale needs to be accomplished for the in-situ material under conditions consistent with the actual construction.

Field testing from an experimental shaft is underway, in conjunction with specialized laboratory testing, to evaluate the swell and reconsolidation behavior.

GENERAL OBSERVATIONS AND CONCLUDING COMMENTS

Earth materials are complex entities, and there is no simple way to classify them with a single term or two. In this paper, a consistent reference frame has been proposed, within which soil and rock masses and their components can be described in a direct, unambiguous manner. Issues related to the overall behavior then are considered, followed by assessment within a particular engineering or construction application. The aggregate result of these three parts answers the question posed by the title of this paper.

The overall implications of this approach are noted, but it is stressed that the overall behavior within the context of a specific set of boundary conditions must always be kept in mind. Extrapolation beyond the defined conditions can be fraught with problems.

Regarding the terminology, we must avoid past practices in which reference is made to a specific entity, such as rock, that does not have a clear and unambiguous definition. We must strive to be precise in our writings and, in particular, in our contract documents. We should perhaps follow the lead of Lewis Carroll and his classic observation from Through the Looking-Glass and What Alice Found There (Carroll, 1960):

> "When *I* use a word", Humpty Dumpty said, in rather a scornful tone, "it means just what I choose it to mean - neither more nor less".

Let's keep our terminology clear, direct, and specific!

ACKNOWLEDGMENTS

Appreciation is expressed to K. J. Stewart and L. Mayes for preparing the text and to A. Avcisoy for drafting the figures.

REFERENCES

Akai, K., Hayashi, M. & Nishimatsu, Y., Eds. (1981). Proc. Intl. Symp. on Weak Rock, Tokyo, 1524 p.
American Geological Institute (1962). Dictionary of Geological Terms, Dolphin Books, Garden City.
Bieniawski, Z.T. (1980). "Rock Classifications: State of the Art & Prospects for Standardization", Research Record 783, Trans. Res. Board, Washington, 2-9.
Canadian Geotechnical Society (1985). Canadian Foundation Engineering Manual, BiTech Publishers, Vancouver, 460 p.
Carroll, L. (1960). The Annotated Alice (Alice's Adventures in Wonderland & Through the Looking-Glass and What Alice Found There), Bramhall House, New York.
Deere, D.U. (1975). "General Rpt.: Applied Rock Mechanics of Weak Materials", Proc. 5th Pan-Am Conf. SM & FE (4), Buenos Aires, 479-

492.

Deere, D.U. & Varde, O.A. (1986). "General Rpt.: Engineering Geological Problems Related to Foundations and Excavations in Weak Rocks", Proc. 5th Intl. Cong. Eng. Geol. (4), Buenos Aires, 2503-2518.

Desai, D.B. & Ku, C.C. (1981). "Mixed Face Tunneling in Urban Setting", Proc. 5th Rapid Exc. & Tun. Conf. (1), San Francisco, 357-382.

Earth Technology Corporation (1990). "Geomechanical Characterization of Eagle Ford Shale at the Superconducting Super Collider Site", Rpt. GR-66, for RTK, Oakland, 28 p.

Flexner, S.E., Ed. (1987). Random House Dictionary of the English Language (Unabridged), 2nd Ed, Random House, New York.

Goodman, R.E. (1976). Methods of Geological Engineering in Discontinuous Rocks, West Publishing, St. Paul, 472 p.

Gove, P.B., Ed. (1966). Webster's Third New Intl. Dictionary of the English Language (Unabridged), Merriam, Chicago.

Hatheway, A.W. (1990). "Weak Rock, Poorly Lithified Cockroaches, & Snakes", AEG News, 33(3), 33-36.

Hudson, J.A. (1989). Rock Mechanics Principles in Engineering Practice, Construction Industry Research & Information Assn./Butterworths, London, 72 p.

Irfan, T.Y. & Cipullo, A. (1990). "Application of Engineering Geological Investigation & Mapping Techniques for Stability Assessment of an Urban Slope in Weathered Rocks in Hong Kong", Proc. 6th Intl. Cong. Eng. Geol. (1), Amsterdam, 141-150.

Kulhawy, F.H., Beech, J.F. & Trautmann, C.H. (1989). "Influence of Geologic Development on Horizontal Stress in Soil", Foundation Engineering: Current Principles & Practices (GSP 22), Ed. F.H. Kulhawy, ASCE, New York, 43-57.

Legget, R.F. & Hatheway, A.W. (1988). Geology & Engineering, 3rd Ed, McGraw-Hill, New York, 613 p.

Leroueil, S. & Vaughan, P.R. (1990). "General & Congruent Effects of Structure in Natural Soils & Weak Rocks", Geotechnique, 40(3), 467-488.

Litke, S. (1991). "An Impressive Feat in Atlanta: Rock Roundtable", Foundation Drilling, 30(1), 31-35.

Meigh, A.C. & Wolski, W. (1979). "Design Parameters for Weak Rocks", Proc. 7th Eur. Conf. SM & FE (5), Brighton, 59-79.

O'Rourke, T.D. & Crespo, E. (1988). "Geotech. Properties of Cemented Volcanic Soil", J. Geotech. Eng., ASCE, 114(10), 1126-1147.

Scott, J.S. (1980). Penguin Dictionary of Civil Engineering, 3rd Ed, Penguin Books, Harmondsworth.

Problems with Rock Excavation Specification In Alberta

Vishnu Diyaljee[1] and Murthy Pariti[2]

Abstract

As a result of contract overruns and claims from classification, measurement, and payment for "solid rock" excavation, Alberta Transportation and Utilities embarked on a dedicated field and laboratory program to investigate and quantify rock materials. This program has resulted in the provision of a rock investigation report for prospective bidders to view, as well as "special provisions" which distinguish rock excavation as common material or solid rock. Since implementation of this program some 3 years ago, no claims have resulted. This paper discusses the contractual problems resulting from rock variability and approaches used in their resolution.

Introduction

The near surface bedrock materials in all parts of Alberta, except the Foothills and Rocky Mountain Belt in the Southwestern part of the Province, consist primarily of marine and non-marine clay shale, siltstone, and sandstone of the Tertiary and Upper Cretaceous ages. For the most part, highway construction activity encounters these material types which are very variable in composition and degree of hardness, and which are considered "soft rock" or rock-like.

Like many highway agencies in North America, the Contract Specifications of Alberta Transportation and Utilities contain clauses covering the methodology to be used in soil and rock construction as well as the measurement and payment for various items associated with these materials.

----

[1]Assistant Director and [2]Senior Geotechnical Engineer
 Alberta Transportation and Utilities, Edmonton
 Alberta, Canada, T6B 2X3

Clause 2.3.03.2A, Section 2.3, of the General and Construction Specifications for Highway and Airport Construction (Contracts Engineering Branch 1987) defines solid rock excavation as follows: "Solid rock shall include rock in solid beds or masses in its original position and boulders or detached rock having a volume of $0.5m^3$ or more". The origin of this specification is unknown to the authors but it is believed that such a specification could have been typically used by many highway agencies in North America through the years. It is likely that this specification could have also been derived from a geological appreciation and description of bedrock.

Over the last 6 years, this specification has received much attention and debate by contractors and Department staff regarding the interpretation and classification of rock, and its measurement and payment. Large contract overruns and attendant claims from four projects in 1985 to 1987 resulted in the Department addressing a dedicated approach to classification, measurement and payment of rock. This effort has utilized rotary drilling, coring, testpit investigation, laboratory identification and classification, point load index testing and unconfined compressive strength testing of rock and rock-like materials.

The purpose of this paper is to discuss the problems that have been associated with rock classification and Contract Administration in highway construction over the last 6 years. The approach used in the last 3 years to minimize these problems both from an investigation and Contract Administration viewpoint will also be addressed.

Bedrock Geology

For an appreciation of the problems associated with rock classification and excavation in Alberta, a review of the near surface bedrock geology is necessary.

A comprehensive picture of the bedrock geology is provided by the Geological Highway Map of Alberta (Geological Highway Map Series 1975) published by the Canadian Society of Petroleum Geologists. The regional geology is illustrated in Fig. 1 which also depicts the major physiographic divisions of the Province.

According to Pawluk and Bayrock (1969), Alberta can be divided into three major geologic "provinces" which correspond to the major physiographic divisions shown in Fig. 1.

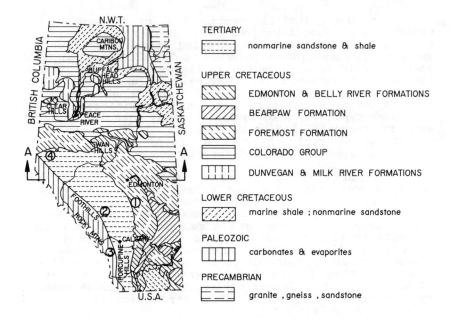

FIG. 1.   Bedrock Geology of Alberta (After Pawluk and Bayrock, 1969)   Section A-A is shown in Fig.2

a.   Canadian Shield   - a small area of Precambrian crystalline rocks in the Northeastern area of the Province adjacent to Lake Athabasca.

b.   Plains Region   - a relative flat area underlain by gently dipping strata of Palaeozoic, Cretaceous and Tertiary ages. This region constitutes most of the Province, and

c.   Rocky Mountain and Foothills   - a band of complexly folded and faulted sedimentary strata extending along the Southwestern margin of the Province.

The Cretaceous and Tertiary Rocks which together underlay all but the Northeastern part of the Plains, constitute a thick succession of interbedded marine and non-marine strata comprised mainly of sandstone and silty shale with subordinate amounts of coal, bentonite, clay-ironstone and volcanic ash.

Some general distinctions in lithology and composition can be made among these rock units but in gross aspect the various sandstone and shales appear similar.

Fig. 2 is a geologic cross-section across the Central and Eastern section (Fig. 1) which represents the majority of near surface bedrock types encountered in highway construction.

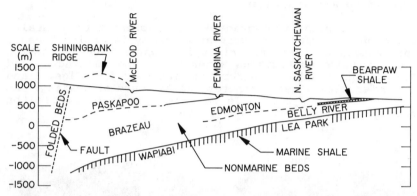

FIG. 2.   **Geologic Section A-A (After Locker,1973)**

The marine beds which outcrop extensively in Northern Alberta adjacent to the Peace and Athabasca River lowlands are dominantly grey shale and siltstone with lesser amounts of sandy detritus. In contrast, much of Central and Southern Alberta is underlain by mixed sandstone-shale formations of non-marine origin, which contain a high proportion of sandy detritus.

The late Cretaceous and Tertiary strata of West Central Alberta are entirely non-marine and appear to differ from the underlying Cretaceous beds mainly in the details of mineral compositions and lithology.

The main composition features of the Cretaceous-Tertiary rocks of the Alberta Plains are the presence of abundant montmorillonite in both sandy and shaly units and the high proportion of unstable volcanic detritus in the sand formation (Pawluk and Bayrock 1969).

Engineering Appreciation of Near-Surface Bedrock

Alberta near-surface rock materials have been generally considered to be, from a geotechnical and general civil engineering perspective,"soft rock" or rock-like.

These rocks are classified as bedrock, although some people might consider bedrock more aptly associated with igneous rock. In general, in Alberta for civil engineering purposes, shales, siltstones and sandstones are accepted as bedrock materials.

The degree of induration of these materials is an important factor in the design of foundations in and within these materials. For example, the shales and siltstones that are poorly indurated are known to weather rapidly when exposed to air and water.

The relative degradation characteristics of shale, siltstone and sandstone were assessed by immersing freshly cut samples of these materials in a water bath and taking time sequence photographs after 5, 10 and 20 minutes and 2.5, 18 and 24 hours (Golder Associates 1986). Figs. 3 and 4 show samples after 5 minutes and 24 hours of soaking. As can be observed, the shale was most degradable while the sandstone was virtually unaffected.

FIG. 3.   After 5 mins soak   FIG. 4.   After 24 hours soak

Other more indurated types can withstand weathering for longer periods and can be very resistant to excavation, requiring in some cases removal by blasting. However, blasting has been mostly associated with hard unweathered sandstone material types.

As a result of the glaciation of Alberta, the near surface bedrock materials are variable in composition, degree of hardness and sequence in stratigraphic profile. It is well known that excavations can encounter hard indurated materials types requiring considerable effort to excavate followed by weathered or weak material at depth.

Based on the understanding of the variability and dispc:ition of near surface bedrock materials, and the glacial geology history of the Province, it is well known to geologists, geotechnical engineers, and the highway construction industry that the bedrock geology influencing highway construction activity is rather complex.

Contract Specifications

The present contract specifications used by Alberta Transportation and Utilities for highway construction identify two principal classes of excavation related to soil and rock. These classes are common excavation in the case of soil, and solid rock in the case of rock. No general definitions are given in the specifications for soil and rock except that for solid rock, a definition is provided as outlined in the "Introduction".

In the engineering literature, the definitions for soil and rock are quite variable depending on whether they are provided by a geologist, civil engineer, engineering geologist or geotechnical engineer. To the geologist the term rock, for example, is applied to all constituents of the earth's crust whether solid (rock), granular (sand and gravel) or earth (such as clay) (Woods et al. 1982).

To the civil engineer and engineering geologist, hard and compact natural materials of the earth's crust are rocks and their derivatives are soils (Krynine and Judd 1957).

The definitions for soil and rock to the geotechnical engineer are best described by Terzaghi and Peck (1967). According to Terzaghi and Peck, the materials that constitute the earth's crust are rather arbitrarily divided by the civil engineer into two categories - soil and rock.

Soil is a natural aggregate of mineral grains that can be separated by such gentle mechanical means as agitation in water. Rock, on the other hand, is a natural aggregate of minerals connected by "strong" and "permanent" cohesive forces. Since the terms "strong" and "permanent" are subject to different interpretations, the boundary between soil and rock is necessarily an arbitrary one.

All rocks can be divided into three general classifications: sedimentary, igneous and metamorphic. These divisions denote primarily the means by which the rocks were formed. As noted, the contract specifications do not distinguish between these rock classifications. Hence, the solid rock specification can be invoked once a material is identified as rock regardless of category.

This specification for rock, if administered according to the intent of its definition, would, perhaps, not be problematic except that there are variable opinions in the construction industry regarding what is or what is not rock.

Rock Investigation and Quantity Determination

Prior to about 1986 the existence of rock on a grading contract was determined principally from the soils survey investigation. Auger drilling was typically used in these investigations and rock was determined from the examination of auger cuttings when drilling became slow and tough and refusal to penetration was generally achieved.

Refusal on sandstone materials, for example, generally resulted in sand particles or cemented sand pieces being obtained from grinding of the auger on this formation.

Rock quantities were determined as a token amount usually 1000 m^3 for projects not containing rock. However, this quantity allowed for a bid price which would be utilized in the event that rock was encountered during construction.

When the actual soil survey indicated the presence of rock, quantities were determined from the roadway cross-section by the highway geometric designer. For many projects, this system proceeded satisfactorily especially where roads were to be widened or twinned and previous information on rock was available. The method of classification, measurement, and payment of rock may have also resulted in the system working satisfactorily.

Contract Administration and Payment

Contract Administration, in relation to rock specification, was somewhat variable throughout the Province. In some cases, payment for rock was made depending on how much effort was required to excavate such materials. If conventional earthmoving equipment including D7 and D8 dozers were capable of moving the rock, then payment was considered as common. In such instances, survey horizons were done to separate materials that could be classified as common from those that were rock.

Payment for a percentage of rock was also used where it was impossible to accurately determine the amount of rock excavation in a cut section. For such situations the percentage of rock in the total cut for that particular cut section was to be specified on the roadway profile or in the contract "special provisions". According to Evjen (1981), this provision was suggested by the Alberta Road Builders and agreed upon by Department Executive.

Problems with Specifications

During the period 1985 to 1988 four contracts were subject to Departmental engineering review by staff of the Contracts Engineering, Design Engineering, Materials Engineering, Regions and Districts. For these contracts, the reviews were primarily related to disagreement between the contractor and Department on classification, measurement and payment for solid rock excavation.

These contracts were associated with the South, Central and Northern sections of the Province and are identified as sites 1, 2, 3, and 4 on Fig. 1. The disagreements were contributing significantly to contract disputes, claims and overruns.

Three of these contracts were designed and tendered prior to 1986 while the other was designed and tendered in 1987.

Contracts for sites 3 and 4 in Southern and Northern Alberta, respectively, were associated with large overruns resulting from the underestimation of design rock quantities. The contract for site 1 in Central Alberta was associated with unbalanced bidding whereby $0.01 was bid for solid rock excavation and the contractor argued that material classified as rock was not solid rock and hence should be paid at the price for common excavation, which was substantially higher.

For site 2 in Southern Alberta, the percentage rock and its payment were challenged through an analysis of the final shrinkage and swell values for the project. According to the contractor, shrinkage values for the project were unrealistic and hence indicative that the percentage of rock was underestimated.

This dispute was resolved by increasing the percentage rock over that which had been agreed upon during various stages of the excavation.

Contracts associated with sites 1 and 3 were subject to evaluation by Golder Associates (1986, 1988) a reputable geotechnical engineering consultant specialized in rock evaluation and assessment. The assessment of the rock materials on both projects was undertaken during or following the completion of excavation of the rock areas. The investigation took the form of an evaluation of the material characteristics through strength testing and fracture spacings.

For both contracts, the consultant agreed that in the strict geological sense the materials encountered were "rock in solid beds... in its original position". However, the consultant also stated that there is an acceptance within the construction industry that several of the bedrock materials in Alberta can be excavated with relative ease, and consequently are either classified or paid for at the same rate as common excavation.

For the contract associated with site 3, there was a disagreement on the definition of the word "solid" and its use in the description of rock masses. To resolve the disagreement it was necessary for the consultant to develop and apply criteria for establishing which bedrock materials were to be paid for at the common excavation rate, and which were to be paid for at the rock excavation rate.

The geotechnical consultant utilized the Kirsten excavatability criteria (Kirsten 1982) as the basis for classification and hence payment. Using the 8 point system of the Kirsten criteria, all materials that were of excavatability class 5 and above were recommended to be paid as rock excavation. This constituted all the siltstones and sandstones while the shale materials were classified as excavatability class 4 and hence recommended to be paid as common excavation.

The U.S. Department of the Navy guide for rock excavation was also utilized in the study and was found to compare favourably with Kirsten's criteria. The U.S. Navy guide, however, does not indicate equipment characteristics as does the Kirsten's criteria. Fig. 5 shows the representation of the shales, siltstones and sandstones on the Navy chart (Golder Associates 1986).

For the contract with unbalanced bidding, a similar study was undertaken by the same geotechnical consultant. For this project, the designed rock quantity was 80,000 m^3. Based on the evaluation, 90,000 m^3 was considered bedrock in the strict geological sense. However, based on the excavatability criteria only, 66,750 m^3 was classified as rock according to the contract definition. The remaining material was classified as common excavation.

Rock Investigation (After 1987)

As a result of the problems associated with rock excavation, classification, measurement and payment during the 1985-1987 period, the geotechnical section of the Department took on the mandate of ensuring that rock was properly classified and quantified.

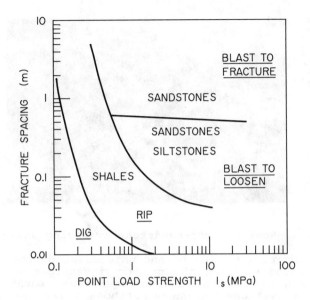

FIG. 5. Suggested Guide For Rock Excavation U.S. Dept of the Navy (After Golder Associates, 1986)

The approach used in the last 3 to 4 years consisted of a dedicated investigation which included test pits, rock drilling and coring, laboratory visual classification, point load index tests and unconfined compressive strength tests. Each project where rock was identified or suspected based on a site review, or review of past information, was subject to a detailed investigation.

The field investigation program consisted of a review of the alignment by the geotechnical engineer and project designer. During this field review, prominent features such as rock outcrops shown in Fig. 6, and deep cuts would be noted. Following this review, a detailed investigation plan would be outlined on the design cross-sections which showed the existing ground elevation and proposed design gradeline.

FIG. 6.   Rock outcrop-sandstone overlying shale

Where rock was to be encountered at depths up to 7m below ground surface, the primary investigation would be undertaken with a Cat 235 or equivalent backhoe. For deeper excavations and where backhoes proved unsuccessful in excavation of test pits, drilling and coring were used as complementary and supplementary methods of investigation.

Based on the field excavation information, laboratory identification and testing, and experience it was determined whether the rock materials encountered would be readily excavated by conventional earthmoving equipment. If a Cat 235 backhoe or equivalent was found to readily excavate the bedrock material (Fig. 7) at successive locations along the route then this material would be classified as common excavation.

For bedrock materials that had to be cored, these were tested for point load index and compressive strengths. These results were used to determine a classification for rock using the rock classification Table 3.4 of the Canadian Foundation Engineering Manual (Canadian Geotechnical Society 1985).

In addition to the dedicated program for field and laboratory investigation, the contract specifications for rock classification, measurement and payment were also addressed. The review of the specifications took into consideration the geological classification of rock as well as the local experience of rock identification, classification and payment.

FIG. 7. **Testpit Investigation**

The Department was particularly concerned about avoiding unbalanced bidding. It was clearly recognized that zero and $0.01 bid prices for rock excavation was an approach used to underbid other contractors at a time when the economic climate was uncertain. These low bidders would also try to prove that rock could be classified and paid as common excavation to achieve some measure of compensation.

Based on the reviews, two categories of rock excavation were agreed upon. One category considered rock being classified as common if conventional earthmoving equipment could be utilized in its removal. For this category, up to D8 dozers were considered as conventional for most contractors. The other category, considered rock to be classified as solid rock excavation whereby extreme effort or blasting would be required to excavate such materials. The classification of this material was guided by visual examination and strength testing of cores obtained from drilling.

For solid rock materials an "extra over" clause for rock quantities was adopted. This methodology was suggested by Hawnt (1988) and accepted by the review committee.

It was felt that this clause would ensure that rock could not be bid at a rate lower than the common excavation rate. For example, a contract with 200,000 m³ of common excavation and 100,000 m³ of rock excavation would be advertised as 300,000 m³ common excavation and 100,000 m³ rock with the price for rock being "extra over" that of the common price. In addition, there was a stipulation that tenders with negative or minus prices would be rejected.

An example of a "special provision" for rock excavation is as follows:

CLASSIFICATION OF ROCK EXCAVATION

1.   General

Further to the preliminary soil survey, the Department has conducted a detailed rock investigation of the cut sections between station 17+220 and station 20+880.

Two copies of this rock investigation report containing testhole logs, photographs of core samples, laboratory test results and cross-section plots are available for viewing at the Office of the District Transportation Engineer. Typical core samples of rock will also be available for viewing.

A large variation of material characteristics was observed in the sandstone formation that was identified as weathered and soft, fractured, and containing thin layers of very hard sandstone. It is possible that similar rock types may be encountered in locations other than those investigated.

2.   Classification, Measurement and Payment

Regardless of what equipment or methods are used, approved excavation will be classified, measured and paid for in accordance with Specifications 2.3.02.4 and 2.3.05, with the exception that the applicable clauses under Section 2.3.02.4.1 and 2.3.05.1.1 for solid rock excavation are modified as follows:

(a) Accepted solid rock excavation in all locations other than the channel excavation locations will be paid for at the contract unit price for common excavation plus the contract unit price for solid rock excavation extra over price of common channel excavation.

(b) Accepted solid rock excavation in channel locations (as defined in Specification 2.3.02.4.2) will be paid for at the contract unit price for channel excavation plus the contract unit price for solid rock excavation extra over price of common or channel excavation.

3. **Quantities**

The estimated quantities for common excavation and channel excavation as shown in the unit price schedule include the anticipated quantities of solid rock excavation.

Summary

Rock materials frequently encountered in highway construction in Alberta are primarily shales, siltstones and sandstones. These materials are sedimentary in origin and are classified as bedrock in the strict geological sense.

The variability of these rock types in terms of quality and spatial distribution have resulted in disagreements between the construction industry and Alberta Transportation and Utilities (contracting agency) in relation to classification, measurement and payment for rock excavation.

From 1985 to 1987 these disagreements led to significant contract overruns and claims. As a result, the Department adopted a dedicated approach to investigating and quantifying rock for contract purposes. This approach resulted in changes to the rock specification by way of special provisions to cater for the rock variability.

As a further attempt to reduce the contract administration problem, a rock investigation report is made available to all bidders at the Department's District office responsible for executing the contract.

Since implementation of these procedures in the last 3 years no claims have so far resulted. While our present approach may be subject to some debate we are nonetheless satisfied that the efforts so far have been rewarding. We are continuing to examine these specifications in light of new information from the construction industry and engineering literature.

Appendix. References

Canadian Geotechnical Society (1985). *Canadian Foundation Engineering Manual, 2nd Edition*. c/o Bitech Publishers Ltd. 801-1030 W. Georgia Street, Vancouver, B.C., p 34.

Contracts Engineering Branch (1987). *General and Construction Specifications for Highway and Airport Construction*. Alberta Transportation and Utilities.

Evjen, P.M. (1981). *Specification Review Committee Recommendations* (Suggestion by the Alberta Road Builders agreed to by R.H. Cronkhite). Internal Memo, Design Engineering Branch, Alberta Transportation and Utilities.

Geological Highway Map Services (1975). *Geological Highway Map of Alberta.* The Canadian Society of Petroleum Geologists, 612 Lougheed Building, Calgary, Alberta, Canada.

Golder Associates (1986). *Final Report to Alberta Transportation and Utilities on the Classification of Materials,* Hwy 22 - km 7.7 to km 29.9.

Golder Associates (1988). *Report to Alberta Transportation and Utilities on the Classification of Materials* - SH 611 Road Cut Between Jct. SH 711 and SH 792.

Hawnt, T. (1988). *Internal memo to V. Diyaljee on SH 520:02 for 3 km west of Lyndon Creek to West of Willow Creek,* File B800J, Alberta Transportation and Utilities, Region 1.

Kirsten, H.A.D. (1982). *A Classification System for Excavation in Natural Materials.* Die Siviele Ingenieur in Suid-Afrika. Julie, 1982.

Krynine, D.P. and Judd, N.R. (1957). *Principles of Engineering Geology and Geotechnics.* McGraw-Hill Book Company, New York.

Locker, J.C (1973). *Petrographic and Engineering Properties of Fine Grained Rocks of Central Alberta,* Research Council of Alberta Bulletin No 30, p 2.

Pawluk, S. and Bayrock, L.A. (1969). *Some characteristics and Physical Properties of Alberta Tills.* Research Council of Alberta, Bulletin No 26.

Terzaghi, K. and Peck, R.F. (1967). *Soils Mechanics in Engineering Practice, 2nd Edition.* John Wiley and Sons Inc.

Woods, F.B., Berry, D.S. and Goetz, W.H. (1982). *Highway Engineering Handbook 1982 Reissue.* McGraw-Hill Book Company.

Acknowledgements

The authors wish to express their appreciation to Swabira Diyaljee and Lavern Millar for their assistance in typing this manuscript and to Les Appleby for drafting the figures.

Where Does Rock Begin Beneath Philadelphia?

Edward F. Glynn[1] and William B. Fergusson[2]

Abstract

Philadelphia provides an excellent example of the engineering problems associated with the interpretation of saprolite/weathered rock/sound rock profiles. The bedrock beneath the city is the Wissahickon Formation of Cambro-Ordovician Age. This formation is a complex of strongly foliated schists intruded by felsic igneous rocks. It grades upward from sound unweathered schist, through a weathered rock zone up to 14 meters thick, into an uppermost saprolite layer as much as 18 meters thick. The rock/weathered rock and weathered rock/saprolite interfaces are transitional, difficult to identify, and are hidden beneath up to 18 meters of unconsolidated clastic sedimentary beds of Cretaceous, Tertiary and Holocene age.

This paper presents a method for identifying the sound rock/weathered rock and weathered rock/saprolite interfaces within the Wissahickon Formation. Identification of these interfaces is accomplished through the assessment of boring log information using rock quality designation (RQD), percent core recovery and standard penetration testing criteria. The paper also discusses the physical characteristics of the sound rock, weathered rock and saprolite and the impact the properties of these materials has had on engineering practice in the Philadelphia area.

- -

[1] Member, American Society of Civil Engineers; Assistant Professor, Villanova University Villanova, PA 19085
[2] Associate Professor, Villanova University Villanova, PA 19085

Introduction

During the past twenty years Philadelphia has experienced a building boom. The downtown area has changed drastically with the completion of new rail and highway links and the addition of several commercial/office developments. The new construction includes projects such as Liberty Place, the Bell Atlantic Tower, the Vine Street Expressway (I-676) and the Center City Commuter Rail Connection. The Rail Connection won ASCE's Outstanding Civil Engineering Achievement Award in 1985.

Geotechnical engineering played a major role in the design and construction of these projects. Many of the foundation and retaining systems required extensive analysis and design. The high rise structures involve column loads much higher than those of earlier buildings. (Prior to 1987 no structure could exceed the height of City Hall - 167 meters.) Foundation design in Philadelphia is complicated by the fact that bedrock is the Wissahickon Formation which grades upward from sound schist and gneiss, through a weathered zone, then through a saprolite layer. The boundaries between the zones are difficult to identify and often tend to be very irregular.

The authors have been developing a geologic data base by collecting boring logs from projects throughout downtown Philadelphia. The inventory currently includes over 700 borings from 70 sites. Not all the borings extend to sound rock. However, the data do reveal the complex nature of the subsurface profile beneath Philadelphia.

This paper summarizes the authors' work to date concerning the relations between raw drill hole data and the geologic and geotechnical aspects of the Wissahickon Formation beneath downtown Philadelphia.

Geology of Philadelphia

Downtown Philadelphia is located slightly north of the confluence of the Schuylkill and Delaware Rivers in southeastern Pennsylvania (Fig. 1). The city lies on both the Coastal Plain and Piedmont physiographic provinces. The downtown area extends 25 city blocks (3.5 km), east to west, from the Delaware River to the Schuylkill River, and 12 blocks (1.9 km), north to south, centered on Market Street. The ground surface in this area is generally between 9 and 12 meters above mean sea level and slopes to an elevation of 6 meters near the two rivers.

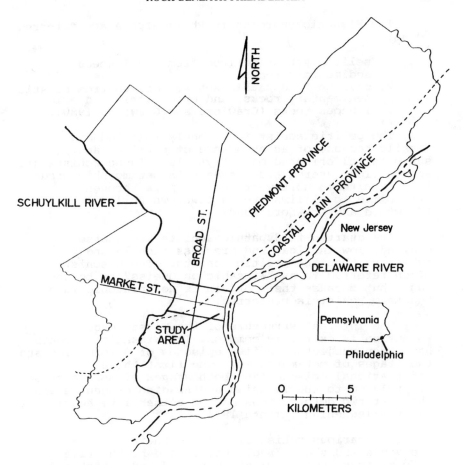

Figure 1. Philadelphia - Physiographic Provinces
(after Paulachok, et al., 1983)

     The bedrock beneath Philadelphia is the
Wissahickon Formation of Cambro-Ordovician age (Higgins
1972). The highly weathered surface of the Wissahickon
slopes eastward at about 10 meters per kilometer. The
surface lies at an elevation of approximately sea level
on the east bank of the Schuylkill River and drops to
an elevation of -30 meters at the west bank of the
Delaware River. The type section of the Wissahickon
Formation is located along Wissahickon Creek about 8 km
northwest of downtown Philadelphia. The closest
outcrop of the formation is about 1.5 km due west of
center city on the west bank of the Schuylkill River.

The Wissahickon rocks in Philadelphia are referred to as:

1. pelitic schist derived from argillaceous sedimentary rocks;
2. quartzo-feldspathic schist derived from clastic sedimentary rocks, and pyroclastic and acid igneous rocks (Crawford and Crawford 1980).

The pelitic schist is strongly foliated and fissile because of an abundance of parallel and sub-parallel oriented mica crystals. These schists are segregation layered and increase in metamorphic grade from kyanite in northwest Philadelphia through staurolite to sillimanite in downtown Philadelphia (Crawford and Crawford 1980).

The quartzo-feldspathic schists are coarse grained, low in mica; and are less fissile and schistose than the pelitic schists. These schists are often referred to as "Wissahickon Gneiss" on boring logs, but because their origin is not plutonic igneous the term schist is preferred.

The internal structure of the Wissahickon Formation is extremely complex. The Wissahickon rocks have been subjected to five episodes of deformation and two stages of metamorphism. The first three deformational episodes and both stages of metamorphism are related to the Taconian and Acadian orogenies and the last two deformations may be either late Acadian or early Alleghenian (Amenta 1974a).

The various schistosities and their associated cleavages and shear fractures, formed by the five deformations, are intricately folded and offset and the earlier deformations are overprinted by later deformations (Amenta 1974b). The rock structure is further complicated by the high grade amphibolite (sillimanite) facies of the second metamorphic stage (Acadian) which overprints the medium grade amphibolite facies of the first metamorphism (Taconian). The rock body is cut by widely spaced (1 to 3 meters), irregular, steeply dipping, open joints. All of these structural discontinuities render the rock mass locally fissile and highly anisotropic.

The joints and open cleavage are arranged in a variety of attitudes into and through the upper portion of the rock body providing a secondary porosity and a groundwater reservoir and a zone of active weathering. This active weathering zone and groundwater reservoir

is a complex network of saprolite, weathered, partly
weathered and sound mica schist.

The Wissahickon Formation is overlain, and
completely covered by unconsolidated beds of
Cretaceous, Pleistocene and Holocene age. Fig. 2 shows
a typical geologic profile in the eastern portion of
the study area. The Cretaceous sediments are absent in
the western portion of the study area. The thickness
of the overlying sedimentary beds in downtown
Philadelphia ranges between 9 and 28 meters. Fig. 3 is
a cross section along Market Street.

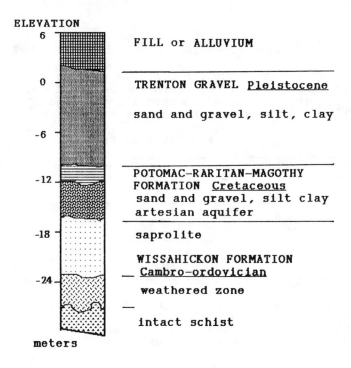

Figure 2. Philadelphia - Geologic Profile

Wissahickon Formation: Engineering Geology.

The interpretations and summaries that follow are
based on a review of the data base and information from
published and unpublished engineering reports. These
data do not provide a definitive picture of subsurface
conditions so the maps and conclusions presented below
must be considered preliminary.

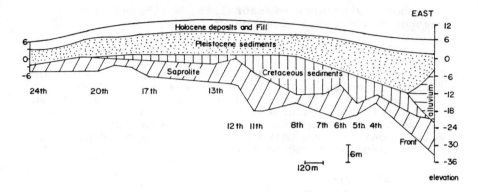

Figure 3. Philadelphia - Cross Section

Table 1 presents the criteria used in this paper to delineate saprolite, weathered rock and sound rock. The criteria are based on core recovery and RQD and are similiar to those proposed by Deere and Patton (1971).

TABLE 1.   Material Designations

Material (1)	Core recovery (%) (2)	RQD (%) (3)
Saprolite	0 - 10	0
Weathered rock	10 - 90	0 - 50
Partly weathered	90 - 100	0 - 75
Sound schist	100	75 - 100

The materials are not so rigorously defined in most boring logs. In many site investigations, top of saprolite is defined as auger refusal and top of rock is defined as tri-cone bit refusal.

Table 2 summarizes some of the properties of the Wissahickon Formation. The values are based on the results of testing programs conducted at a site on Market Street approximately 1.0 km east of Broad Street (Fergusson and Glynn 1988).

The strength parameters were measured in consolidated drained triaxial tests. The orientation of the schistosity planes affects the strength of the samples. The samples tend to fail along surfaces inclined at small angles to the schistosity planes. Partos et al. (1989) documented the relation between

TABLE 2.   Typical Properties

Material (1)	w (%) (2)	d (kN/m³) (3)	q (kPa) (4)	φ (°) (5)	c (kPa) (6)
Saprolite	18	18	150	22	135
Weathered rock	14	19	200	22	240
Partly weathered	7	21	1300	18	700
Sound schist	1	26	15000	27	2100

w: water content
d: unit weight of material
q: unconfined compressive strength
φ: effective friction angle
c: effective cohesion

the strength parameters and schistosity angle in their paper on the Liberty Place foundations.

Saprolite   The saprolite was formed in place by intense chemical weathering by circulating groundwater on the exposed Wissahickon Formation surface.   Saprolite development started some time after the Triassic and continues to the present.

The saprolitic material is a soft, friable, mixture of silt, sand, and gravel size particles whose composition, structure and fabric is dependent upon the parent rock.   The Unified Soil Classification System designations are typically GM or SM.   The thickness, structure, texture, composition and physical characteristics of the saprolite vary greatly both horizontally and vertically.

The standard penetration resistance (N) of the saprolite increases with depth.   The average N is around 100 blows.   However, the presence of core stones in the saprolites formed from igneous rock, and quartz fragments from the schists, cause the N values to be misleading.

The average saprolite thickness in the downtown area ranges between 3 and 12 meters as shown in Fig. 4. The saprolite is thickest in the center of downtown and thins east towards the Delaware River and west towards the Schuylkill River.   The contours show the general trend of the saprolite thickness and do not reflect local variations which may be severe.

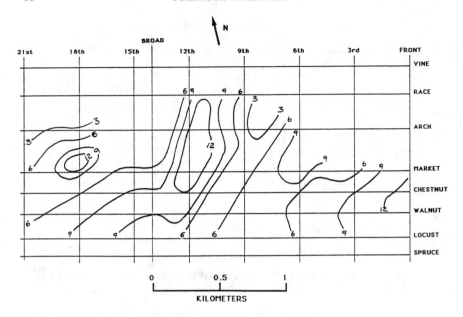

Figure 4. Downtown Philadelphia - Average Saprolite
Thickness - Thickness Contour Interval 3
Meters

     The greatest saprolite thicknesses are usually
found to the east of the center of the city where they
have been protected from erosion by the overlying
Cretaceous, Pleistocene and Holocene flood plain
deposits. The saprolite has been thinned and locally
removed by erosion beneath the gravelly channel
deposits of Cretaceous age and recent erosion by the
Delaware and Schuylkill Rivers.

     The local variation in saprolite thickness is
illustrated in Fig. 5. This site measures 50 meters by
50 meters and is located near the intersection of Arch
Street and Broad Street. The site investigation
included 18 core borings. The logs indicate that the
saprolite is generally about 4 meters thick. However,
there is a zone near the northern boundary where the
thickness reaches almost 11 meters.

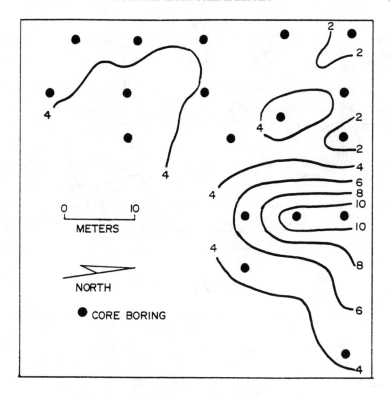

Figure 5. Downtown Site - Average Saprolite Thickness -
Thickness Contour Interval 2 Meters

Weathered and Partly Weathered Rock   The interface
between the saprolite and the underlying Wissahickon
Formation (Fig. 6) contains a zone of weathered rock
whose thickness varies from less than a meter to as
much as 12 meters.   The top of the zone generally dips
to the east; however, there are two depressions along
Market Street at 6th and at 18th Streets.

        Local variations are superimposed on the general
trends noted in Fig. 6.   Fig. 7 shows in the contours
of the saprolite/weathered rock interface at the site
described in Fig. 5.   In many areas of the site the
interface drops nine meters over a distance less than
eight meters.   Some of these local depressions may be
the remnants of ancient streams.

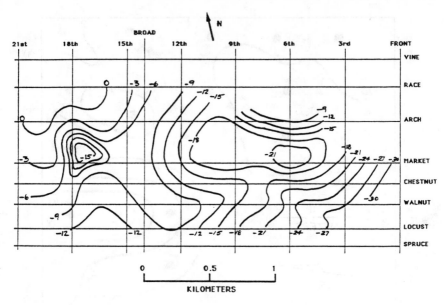

Figure 6. Downtown Philadelphia - Contour Map of Rock
Surface - Contour Interval 3 Meters

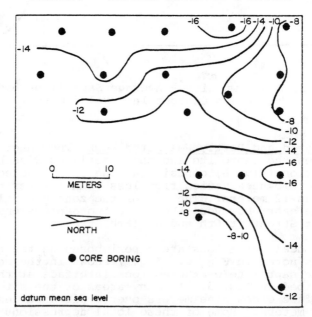

Figure 7. Downtown Site - Contour Map of Rock Surface -
Contour Interval 2 Meters

The weathered zone is an active area of saprolite formation. There is usually a steady downward increase in RQD but often the degree of weathering varies vertically and frequently weathered and partly weathered zones are found within the saprolite. This occurs because the weathering process constantly attacks the minerals that are susceptible to weathering, such as the feldspars, amphibole and biotite, which are segregated into definite zones within the schist. The segregation zones of easily weathered minerals cause selected zones of weathering that are dependent upon the availability of circulating groundwater following the steeply dipping schistosity.

Sound Mica Schist Typical properties of sound mica schist are noted in Table 2. The unconfined compressive strength varies over a wide range. Strengths as high as 42,000 kPa have been reported. The higher strength samples often contain quartz zones and the foliation is parallel or subparallel to the vertical axis of the core sample.

In the Philadelphia area sound rock is typically defined as material with an unconfined compressive strength greater than 15,300 kPa. Very few core borings extend into material that meets the core recovery/RQD criteria for geologically sound rock presented in Table 1.

Deep Foundations in Downtown Philadelphia

Most of the high-rise structures in downtown Philadelphia are supported on drilled piers. Table 3 summarizes some of the foundation details for 4 notable structures, all of which are founded on the Wissahickon.

TABLE 3. Drilled Pier Foundations - Philadelphia

Building (1)	Date (2)	Height* (3)	Type of Pier (4)	BRP** (kPa) (5)
PSFS	1932	39	belled	1900
5 Penn Plaza	1970	36	belled	2850
Liberty Place	1986	61	socketed	3800
Independence Sq	1981	17	socketed	5700

  * Height of building in stories
 ** Design bearing pressure for drilled piers

As noted in the table, some of the newer structures have the piers socketed in rock. A shaft resistance or rock adhesion of 200 to 900 kPa is typically used in design.

Most of the recent designs are based on a 3800 kPa bearing pressure for material with an unconfined compressive strength of at least 15,300 kPa. Most job specifications do not require a laboratory test at each drilled pier. They require the contractor to take rock samples from the bottom of selected piers. The samples, typically 50 mm diameter by 100 mm long, are tested in unconfined compression. The strength of each sample is then compared to the penetration rate of a standard pneumatic drill at the same piers. The purpose of the study is to establish a maximum penetration rate for 15,300 kPa material. The maximum penetration rate, i.e., minimum time to drill 1.5 meters, serves as a criterion to evaluate the founding grade of drilled piers which have no laboratory tests.

Conclusions

The expected test boring sequence of saprolite, weathered rock, partly weathered rock, sound rock seldom occurs in such an orderly sequence. More often the weathered zone contains intertonguing layers of saprolite, weathered, partly weathered and sound rock between the main saprolite deposit above and the sound rock below.

The weathered, partly weathered and sound rock zones are intercalated within the sound rock where groundwater from the saprolite/weathered rock interface follows readily weathered minerals along the fractures and foliation of the schist. The weathered zones and the contacts between them are gradational.

The material designation - sound rock - as defined in Table 1 is a geological interpretation and denotes a fresh, unweathered rock specimen. Sound rock is rarely encountered in the core borings because the partly weathered rock near the base of the weathered zone of the Wissahickon Formation has sufficient strength to accommodate foundation loads.

APPENDIX I. REFERENCES

Amenta, R. V. (1974a). "Relative timing between
deformation and metamorphism in the Wissahickon
Formation near Philadelphia." Guidebook for the
39th Annual Field Conference of Pennsylvania
Geologists - Geology of the Piedmont of Southeastern
Pennsylvania, King of Prussia, PA, 10-13.

Amenta, R. V. (1974b). "Multiple deformation and
metamorphism from structural analysis in the eastern
Pennsylvania Piedmont." Geological Society of
America Bulletin, 85, 1647-1660.

Crawford, M. L., and Crawford, W. A. (1980).
"Metamorphic history of the Pennsylvania Piedmont."
Journal of the Geological Society, 137, 311-320.

Deere, D. U. and Patton, F. D. (1971). "Slope stability
in residual soils." Proc. 4th Panamerican Conference
on Soil Mechanics and Foundation Engineering, San
Juan, PR, 87-170.

Fergusson, W. B. and Glynn, E. F. (1988). "A foundation
failure in Philadelphia." Proc. 2nd Internation
Conference on Case Histories in Geotechnical
Engineering, St. Louis, MO, 1293-1296.

Higgins, M. W. (1972). "Age, origin, regional
relations, and nomenclature of the Glenarm Series,
central Appalacian Piedmont: a reinterpretation."
Geological Society of America Bulletin, 83,
989-1026.

Partos, A. J., Sander, E. J., and Hungspruke, U.
(1989). "Performance of a large diameter drilled
pier." Proc. International Conference on Piling and
Deep Foundations, London, 309-316.

Paulachok, G. N., Wood, C. R., and Norton, L. J.
(1983). "Hydrologic data for aquifers in
Philadelphia, Pennsylvania." Open File Report
83-149, U.S. Geological Survey.

Special Bedrock Conditions in Greater Boston

John T. Humphrey[1] , M. BSCE, AIPG, AEG
and
Cetin Soydemir[2] , F. ASCE

Abstract

The Cambridge Argillite, a weakly metamorphosed shale, forms the major portion of the Boston Basin, a structural as well as topographic depression. In the greater Boston area, the argillite bedrock and/or over- lying glacial till (hardpan) provide suitable support for a wide range of foundation systems. The bedrock is generally hard and competent due to its poorly developed bedding planes and lack of fissility. However, locally and in an irregular manner, hydrothermal alteration of the bedrock has created zones of highly decomposed, kao- linitic materials of soft consistency and medium plas- ticity that vary in thickness from a few to hundreds of feet. Generally lying at depth, beneath high water table conditions, the altered argillite has been incor- rectly identified as the overlying till or, in some instances, the famous Boston Blue Clay. The zones of highly altered bedrock present major geotechnical con- cerns, and their unpredictable occurrence and orienta- tion will have a considerable impact on foundation designs and construction in the Boston area.

Regional Geology

Boston, Massachusetts lies in the Boston Basin, a portion of the Boston Lowland Geomorphic District of the New England Physiographic Province. The Boston Basin consists of a very complex downwarp, or structural

---

[1] Chief Geologist, Haley & Aldrich, Inc., 58 Charles Street, Cambridge, MA 02141.

[2] Vice President, Haley & Aldrich, Inc., 58 Charles Street, Cambridge, MA 02141.

synclinorium, made up of rocks called the Boston Bay
Group, dating from the late Precambrian (Proterozoic Z)
geologic period. The Basin is broken into several east-
northeast striking anticlines and synclines that are
separated by major inactive faults. Zones of tight
folding and crushing mark the axes of the major folds.

The Boston Bay Group has traditionally been
divided into two principal formations: the Roxbury Conglomerate and the Cambridge Argillite (Figure 1). The
thickness of the Boston Bay Group is estimated to be in
excess of 17,000 ft (5200 m), of which the Cambridge
Argillite makes up approximately 4,000 ft (1200 m).
The bedrock underlies the Basin at depths varying from
about 20 ft (6 m) to 200 ft (61 m) below ground surface. The metasedimentary and volcanic units have been
randomly intruded by younger igneous dikes and sills of
varying compositions.

The bedrock is generally hard and competent due
to its poorly developed bedding planes and lack of
fissility. However, localized areas exist where alteration of the bedrock has produced zones of varying
widths of kaolin, a rock composed essentially of the
clay mineral kaolinite. The geologic processes that
have created these "soil-like" zones have been a subject of ongoing controversy (Kaye, 1967; Kaye and
Zartman, 1980; Kaye, 1982). The origin of the highly
altered kaolinitic zones is of practical importance to
construction engineers. If the alteration is due to
geologically "recent" mechanical weathering processes,
the kaolin would be assumed to be a relatively thin,
but continuous surficial zone overlying fresh, sound
bedrock. If, however, it is the remnant of deep, long-
term chemical weathering, the kaolin could occur intermittently and extend to unpredictable depths. If the
alteration is associated with younger Triassic-age
igneous dikes and sills that have intruded the bedrock,
the adjacent hydrothermally altered zones could be
random in their direction, thickness and depth.

The Boston Basin is a topographic as well as a
structural depression and was subjected to continental
glaciation during the Pleistocene Epoch. The Basin was
filled with glacial and post-glacial marine and estu-
arine sediments. A representative geologic profile in
the greater Boston area may consist of (from ground
surface down): fill, organic silt, silty sand, silty
clay (Boston Blue Clay), sand and gravel (glacial
outwash), glacial till, and bedrock (Figure 2). The
glacial till, known locally as "hardpan," typically
ranges from several feet to several tens of feet in
thickness and is composed primarily of sandy silt to

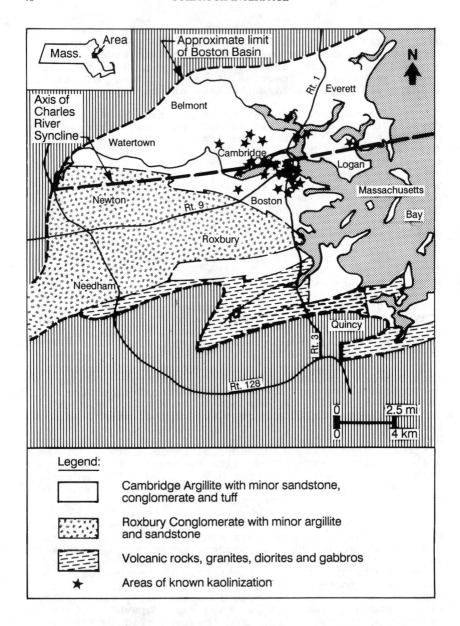

Figure 1. Generalized Geologic Map of the Boston Basin.

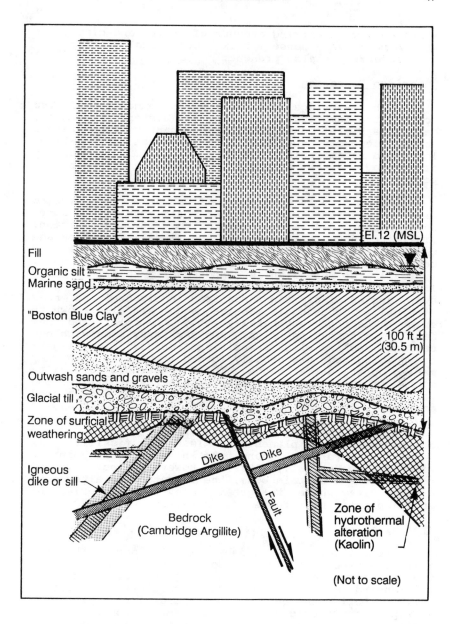

Figure 2. Schematic Geologic Conditions beneath Boston.

silty sand with varying amounts of gravel and clay in a
well-bonded, very dense matrix. Occasional cobbles
and boulders are also present.

Cambridge Formation

The Cambridge Formation, more commonly referred
to as the Cambridge Argillite, and historically as the
Cambridge Slate, consists of predominantly fine-grained
marine deposits that have been subjected to long-term,
low-grade metamorphism, and then intruded by a variety
of igneous dikes and sills during the Triassic time
period.

Argillite is generally defined as a rock formed
from a claystone, siltstone, or shale, but with a somewhat higher degree of induration. By definition,
argillite does not exhibit the fissility or cleavage
characteristics of shale or slate. However, localized
occurrences of fissility are known to exist in the unit.
Typically the formation is composed of steeply dipping
(50° to 60°), relatively thinly bedded units. However, angles of dip can range from horizontal to near
vertical. Occasional massive beds are also known to
exist. The rock is generally gray in color and may
display alternating light and dark gray beds with occasional red and black transitional units. The light-
er colored beds are usually formed from silt, with clay
making up the darker bands. Thick sequences of siltstone or sandstone may occur within the argillite, and
volcanic tuff (consolidated ash) may be encountered
anywhere within the formation.

The Cambridge Formation, due to its poorly
developed bedding planes and lack of fissility, is
generally hard and sound from an engineering point of
view (Cullen et al., 1982). However, locally, and
restricted to certain beds, the formation is altered to
a soft, dirty-white to green, clay-like material of
medium plasticity referred to as kaolin.

Kaolinization

In 1914, W.O. Crosby of MIT observed that the
argillite exposed in the foundation excavation for the
new Cambridge site of MIT was highly decomposed "to a
whitish and more or less plastic clay" (Worcester,
1914). During the period 1950 through 1960, the Boston
Metropolitan District Commission (MDC) constructed a
series of rock tunnels throughout the Boston Basin for
sewer and water transmission. M.P. Billings of Harvard
University and his associates mapped the tunnel excavations and published a series of reports and maps (Rahm,

1962; Billings and Tierney, 1964; Billings and Rahm, 1966; Tierney et al., 1968). An interesting fact that is common to all these reports is the amount of kaolinization encountered in both the argillite and the much harder conglomerate. A 40-ft (12 m) thick kaolin layer was penetrated in the Boston City Tunnel at about 300 ft (92 m) below the rock surface. However, when the Geological Society of America (GSA) published Billings' definitive "Geology of the Boston Basin," which summarizes all of the various tunnel mapping efforts, there is no mention of kaolinization or decomposed bedrock (Billings, 1976). Other investigators have also expressed doubts whether this material is in fact bedrock (Lambe, 1960).

With the recent proliferation of subsurface explorations in the greater Boston area associated with major engineering projects, it has been confirmed that highly altered bedrock not only occurs in discrete and isolated beds, but will have a considerable impact on deep foundation projects presently being considered in the greater Boston area.

The geochemical mechanisms for the high degree of bedrock alteration (kaolinization) is subject to various interpretations. However, it now appears that the original marine deposits were mixed with, and intermittently covered by, thin eruptive volcanic ash units that altered the immediately adjacent material. Continued deposition, followed by additional eruptions, created a "sandwich" effect of sound and altered bedrock units. Subsequent folding and faulting of the Basin rocks tilted the original horizontal east-west striking beds into relatively high angles. Long-term erosion and geologically more recent glacial action have exposed the ends of these beds. At many locations throughout the Basin, fresh unaltered strata are found interbedded with highly altered rock, and both types alternate irregularly with rocks showing intermediate degrees of alteration (Figure 3).

Kaolinization has been encountered and reported at many places in the Boston Basin (Figure 1). The alteration has been found in all rock types, including the conglomerate. Thickness of these clay-like "soil" zones may vary from a few inches to hundreds of feet. Studies of thin sections from the altered zones reveal that the minerals normally present in the argillite, including quartz, have been variously replaced by the minerals sericite and kaolinite. Kaye (1967) also noted that isolated younger igneous dikes and sills have been variously altered to siderite (iron carbonate) and minor kaolinite.

Scale in (ft)	Core Run Number	REC / RQD	Field Classification
	C1	60 / 95	Hard, gray ARGILLITE, moderately fractured, high angle tight joints. 119.0 ft
–120			Medium hard, gray-green TUFF, highly fractured and jointed.
	C2	40 / 50	124.0 ft
–125			Soft, light gray to white, highly decomposed ARGILLITE (soil like).
	C3	20 / 0	129.0 ft
–130			Hard, black, highly fractured DIABASE (dike). 132.0 ft
	C4	60 / 85	Medium hard, gray ARGILLITE, slightly fractured.
–135			

REC: Percent recovery from 5 ft. NX core run.
RQD: Percent rock quality designation.

Figure 3. Excerpt of Test Boring Log of Bedrock,
Downtown Boston.

It appears that alteration of the argillite to kaolin has not taken place adjacent to all of the numerous and various igneous dikes and tuffaceous units that occur throughout the Boston Basin. The abrupt and unpredictable change from the sound argillite to the kaolinitic, weak argillite may occur in a very short distance. To further complicate the situation, pre-glacial surficial weathering has created a variable thickness of the overlying weathered rock that appears to have characteristics similar to those of the kaolin-ized zones. This surficial zone of mechanical weather-ing is generally not continuous, having been locally removed by subsequent glacial action.

Test borings in the downtown Boston area have penetrated up to 300 ft (92 m) of the soft kaolinitic zone, without encountering the sound, competent argil-lite that may be only a few meters away in a horizontal direction (Kaye, 1967). Other investigators have indicated kaolinitic zones extending to even greater depths.

Geotechnical Engineering Considerations

In the greater Boston area, glacial till and bedrock generally provide suitable support for a wide range of foundation systems, including high capacity end-bearing piles. However, in some areas where the glacial till is absent or relatively thin, and the underlying rock is highly altered, geotechnical engi-neers are faced with special foundation design prob-lems. The unaltered, unweathered Cambridge Formation consists of relatively hard, brittle and elastic material. The localized decomposed or faulted zones may be difficult to identify with randomly spaced test borings. Velocity measurements obtained from geo-physical methods generally cannot distinguish between the overlying glacial till and the underlying kaolin-ized rock, if it is present. Engineering properties determined for the bedrock in field and laboratory tests must be evaluated with care and judgment.

As bedrock is relatively deep in the downtown Boston area, there is a paucity of information regard-ing its engineering properties. Limited data on the hard fresh argillite indicates an average unit weight of 170 pcf (2,723 kg/m^3), with an average unconfined compression value of 18,000 psi (124 N/m^2). Depend-ing on the lithology of the unit, variations in these values occur both well below and slightly above the average. Laboratory tests conducted on the highly altered kaolin reveal an average liquid limit of 35, a plastic limit of 25, and a plasticity index of 8.

These values classify the highly altered argillite bedrock as a clay.

High capacity end-bearing piles designed to support relatively high loads, while being driven to the required penetration resistance, may penetrate through the glacial till layer where it is relatively thin and advance into the much less resistant, highly altered argillite to depths greater than anticipated during the design phase. In one project (Soydemir and Humphrey, 1989), this situation required the use of piles 20 ft (6 m) to 30 ft (9 m) longer than originally planned (Figure 4). An alternative option is to reduce

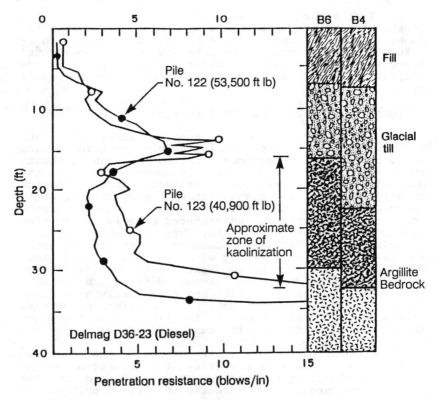

Figure 4. Typical Pile Driving Performance in Altered
Argillite (Kaolin).

the design capacity of the piles so that the necessary end-bearing support may be mobilized within the relatively thin glacial till stratum. Driving additional but shorter piles may in some cases be a more economical solution.

A variety of problems may occur if a relatively thick unit of unanticipated soft kaolin were encountered in a tunnel excavation, whether it was excavated by TBM or drill-and-blast methods. In addition, the igneous dikes and sills, which randomly criss-cross the Boston Bay Group, are generally fractured and provide easy access for groundwater in zones considered "impermeable."

Straight shaft and bell caissons as well as slurry (diaphragm) wall installations to bedrock may encounter varying bedrock conditions that could severely impact both their design and installation, as illustrated in Figure 5.

In a recent project, test borings obtained for foundation design penetrated 10 ft (3 m) into "sound" bedrock. The design called for a drill-and-blast excavation of 15 ft (5 m) into the rock. The bedrock was "shot" and no unusual conditions were described by the drilling subcontractor. Before the prime contractor was aware of any problems, his earthwork subcontractor had excavated from 5 to 10 ft (1.5 to 3 m) below the design grade in one area of the site. As the earthwork subcontractor had been instructed to remove all broken and fractured rock, the drilling subcontractor was held responsible for overblasting. Upon examining the site, it was immediately apparent that a 12-ft (4 m) thick, gently dipping, sound sandstone unit overlay a thick altered argillite unit. The 2-cu yd excavator had penetrated the previously blasted "cap rock" and the underlying soft kaolin with exceptional ease.

Summary and Conclusions

It is well established that the localized kaolinitic zones in the Cambridge Formation present major geotechnical design and construction concerns, and that their unpredictable occurrence will have a considerable impact on the deep underground projects presently being considered in the greater Boston area.

In order to minimize these impacts, detailed and extensive test borings and in-situ testing should be mandatory during both design and construction phases of the work. Early identification of the potential problems is possible using experienced field geologists

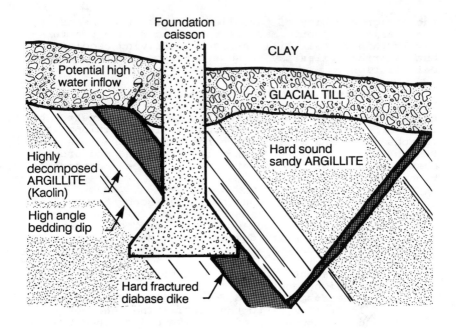

Figure 5. Typical Bedrock Foundation Conditions,
           Cambridge Formation, Boston.

knowledgeable   about   local subsurface conditions, teamed
with qualified drillers equipped with a   wide   range   of
drilling   and   sampling equipment.   It is all too common
to "drill ahead" in the decomposed and   altered   bedrock
in   order   to seat the   drill casing on sound bedrock,
prior to coring.   In many instances, however,   the   zone
penetrated without any   samples is always suspect and
may well be   erroneously   identified   as   the   overlying
glacial till.

        If very large loads of proposed structures are
to be supported on caissons   or   load bearing elements
associated with new innovative "up-down" construction
techniques, it is recommended that triple tube rock

core boring techniques be used at each location. These
procedures will insure maximum rock core recovery for
properly assessing if a stratum of soft, highly altered
rock occurs at a specific location, or if a permeable,
fractured igneous dike is present. Specialty in-situ
testing using pressuremeters or dilatometers will
further help in isolating these problem zones.

It appears that, at least for the present,
identification of the soft kaolinized soil/rock zones
within the competent bedrock will be extremely
difficult at best and their occurrence and distribution
remain unpredictable. It may be necessary for the
Owner, Engineer, and the Contractor to agree, prior to
commencement of the project, to share the inherent
financial risk associated with a kaolin encounter!

Appendix 1. References

Billings, M.P. (1976). "Geology of the Boston Basin." Geological Society of America, Memoir 146.

Billings, M.P., and Rahm, D.A. (1966). Geology of the Malden Tunnel, Massachusetts, Boston Society of Civil Engineers Journal, v.53, No. 2, 116-141.

Billings, M.P., and Tierney, F.L. (1964). Geology of the City Tunnel Extension, Greater Boston, Massachusetts, Boston Society of Civil Engineers Journal, v.51, No. 2, 111-154.

Cullen, T.R., et al. (1982). "Tunneling through the Cambridge Argillite." Proceedings, Geotechnology in Massachusetts, University of Massachusetts, 197-211.

Kaye, C.A. (1967). "Kaolinization of Bedrock of the Boston, Massachusetts Area." Geological Survey Professional Paper 575.

Kaye, C.A. (1982). Bedrock and Quaternay Geology of the Boston area, Massachusetts, Geological Society of America, Reviews in Engineering Geology, v.5, 25-40.

Kaye, C.A., and Zartman, R.E. (1980). "A Late Proterozoic Z to Cambrian Age for the Stratified Rocks of the Boston Basin, Massachusetts, USA." in Wones, D.R. ed., The Caledonides in the USA, Virginia Polytechnic Institute and State University, Dept. Geological Science Mem. 2, 257-264.

Lambe, T.W. (1960). "Subsoils at the Site of the Earth Science Building at Massachusetts Institute of Technology, Cambridge, MA." MIT Soils Engineering Department Report (duplicated).

Rahm, D.A. (1962). Geology of the Main Drainage Tunnel, Boston, Massachusetts, Boston Society of Civil Engineers Journal, v.49, 319-368.

Soydemir, C., and Humphrey, J.T. (1989). "Special Bedrock Conditions and a Foundation Project in Greater Boston." Proceedings, Foundation Engineering Congress, Evanston, Illinois, v.2, 1420-1431.

Tierney, F.L., et. al. (1968). Geology of the City Tunnel, Greater Boston, Massachusetts, Boston Society of Civil Engineers Journal, v.55, 60-96.

Worcester, J.R. (1914). Boston Foundations, Journal of the Boston Society of Civil Engineers, v.1, 223-232.

WHERE DOES ROCK BEGIN FOR
THREE HIGHWAY PROJECTS ?

Glenn M. Boyce, A.M. ASCE [1] and Lee W. Abramson M. ASCE [2]

ABSTRACT: "Where does rock begin?" was a question raised on three large highway projects in the western United States.   The question was successfully answered by a variety of exploration approaches. Each of these case histories is discussed in detail.   Attention is provided to the nature of the problem, what methods were used to determine where rock began, how successful the various exploration and testing methods were, and how the problems were solved.

INTRODUCTION

"Where does rock begin?" is a question raised on many design projects. Normally the question concerns the interface between soft overburden soil and hard bedrock for excavations and foundation support. A twist to the question arose on three large highway projects in the western United States. On these projects, the question "Where Does Rock Begin?" involved determining where talus contacted rock, and where highly weathered rock (saprolite) contacted less weathered rock.   The three projects were the Interstate Route H-3 Tunnel Project on the Island of Oahu in Hawaii, the Provo Canyon U.S. Route 189 Tunnel Project in Utah, and the Glenwood Canyon Interstate 70 Tunnel Project in Colorado. The question was successfully answered by a variety of exploration approaches as described below. Disputes Review Boards and Design Summary Reports were also incorporated into the contract documents and helped avoid problems associated with determining where rock begins.

INTERSTATE ROUTE H-3 TUNNEL PROJECT

The first project was the Interstate Route H-3 Highway on the Island of Oahu, Hawaii. This project involved the planning, design, and construction of a 10-mile (16-km) long final segment of new highway traversing the Koolau

---

[1] Senior Project Engineer, Parsons Brinckerhoff Quade & Douglas, 303 2nd St., Suite 700N, San Francisco, CA 94107-1317
[2] Manager, Underground & Tunnels Group, Parsons Brinckerhoff Quade & Douglas, 303 2nd St., Suite 700 N, San Francisco, CA 94107-1317

57

Mountain Range. The highway connects Kaneohe on the windward side with Honolulu on the leeward side of the Island as shown in Figure 1. The project included two 1-mile (1.6-km) long tunnels through the mountain range made up of basalt in different degrees of weathering states. The twin-bore Trans-Koolau Tunnels will have two lanes in each direction and a roadway width of 38 feet (11.6 meters) including shoulders. Construction began on the project in late 1987 and is expected to finish in late 1994.

The geology consists of lava flows originating from the ocean floor, forming the island chain. Lava is expelled from volcanos and deposited in layers. These basalt flows consist primarily of four types: pahoehoe, transitional, aa, and clinker. The types of flow depend on the velocity the lava travels. All flow types are permeable because of the combination of voids, horizontal differential shearing joints, and vertical cooling fractures. These four flow types form successive layers and become interstratified. Because of the permeable nature of the flow types, water infiltrates the rock mass, weathering the materials. Eventually the massive rock decomposes and becomes almost soil-like. Highly to extremely weathered rock is known as saprolite. Saprolite acts like a clayey silt material when disturbed. It looses much of its strength when drilled or excavated.

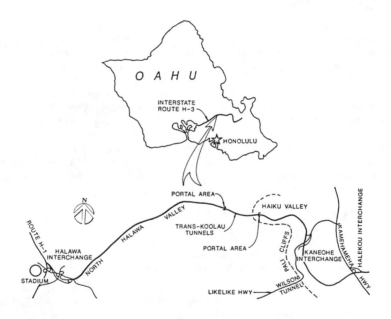

Figure 1. Location Map of the Interstate Route H-3 Highway Project on the Island of Oahu.

Degrees of weathering were defined for exploration and geologic mapping purposes. The definitions were based on the physical disintegration due to chemical alteration of the minerals in the rock. Terms and abbreviations used to describe weathering are:

EW Extremely Weathered - The original minerals of the rock have been almost entirely altered to secondary minerals, even though the original fabric may be intact. Also termed saprolite.

HW Highly Weathered - The rock is weakened to such an extent that a 2-inch (51-mm) diameter core can be broken readily by hand across the rock fabric. Also termed saprolite.

MW Moderately Weathered - Rock is discolored and noticeably weakened, but a 2-inch diameter core cannot usually be broken by hand across the rock fabric.

SW Slightly Weathered - Rock is slightly discolored, but not noticeably lower in strength than fresh rock.

UW Unweathered - Rock shows no discoloration, loss of strength, or any other effect of weathering. Also termed fresh rock.

Design Problem. The weathering process on the Island has created a fairly uniform geologic profile. The surface is covered with a residual soil. The composition of the material below the surface is extremely weathered. With depth, the material begins to retain more of the rock's fabric characteristics. At greater depths, the material is rock-like with only the joints weathered and filled with softer material. This generalized profile is common and does not hamper conventional construction procedures (foundations, cuts, and fills). In contrast, tunnels in this mountainous setting intercept the soil, weathered rock, and rock. Construction of the Wilson Tunnels on the Island of Oahu (Peck, 1981) dramatically illustrated the effects of construction through transitional ground.

The Wilson Tunnels are part of the Likelike Highway that traverses the same Koolau Mountains approximately 2.5 miles (4 km) south of the H-3 alignment. In 1956 during construction of the first Wilson Tunnel, a collapse occurred, killing a dozen miners. The collapse occurred when groundwater was encountered at the soil/rock interface as the tunnel was driven full-face from the rock into the soil. The soil consisted of saturated decomposed rock and was sensitive when disturbed. Encountering this weak material in its saturated condition initiated a collapse which rose to the surface forming a sinkhole. An exploration program followed by the construction of an exploratory tunnel was initiated to safeguard against a similar failure in the new Interstate H-3 Tunnels. Within the exploratory tunnel, an extensive geotechnical testing program of plate load and pressuremeter testing was conducted to identify the soil/rock interface and tunneling conditions through the weathered rock.

First Phase Exploration Program. Geologic information for the project was needed prior to conventional access to the site. Vertical and inclined borings were drilled with the use of helicopters for hauling the drill rigs, workers, and materials to the boring locations. Foam drilling techniques were used to conserve drilling water, which had to be flown to the drill sites in barrels. Special drilling and logging methods were developed for the project to characterize the weathered saprolite and basalt (Abramson and Hansmire, 1989). Rock quality was characterized by Rock Quality Designation (RQD). Because of the weathered condition of the rock, a modification of RQD suggested by Deere and Deere (1988) was employed. Rock in an unweathered (UW) or slightly weathered (SW) condition was counted in RQD in the traditional manner. In the modified system, moderately weathered (MW) rock was also counted and the designation was flagged by an "*", e.g., RQD*.

The borings in Figure 2 showed less weathered material than expected at the Halawa Portal. Extremely weathered material (designated with the symbol "E") was expected based on the Wilson Tunnel experience. Initially, the core was classified as moderately to unweathered rock.

Exploratory Tunnel. An exploratory tunnel was planned and constructed to locate any trapped dike water within the mountain and to locate the boundary or contact between soil (weak, highly weathered rock) and rock. To determine where rock began, an extensive in situ geotechnical testing program that included pressuremeter and plate load tests, laboratory tests, and convergence measurements was completed (Boyce and Abramson, 1991).

The main exploratory tunnel was located in the pillar between the two future main tunnels to act as a drainage gallery in the event that large quantities of trapped dike water were encountered. There were also drifts in the crowns of the future main tunnels that ramped up from the main exploratory tunnel. Figure 3 shows the configuration of the ramp, crown drifts, and geotechnical test locations. The crown drifts were driven from within the mountain, outwards toward the future highway portals. This direction of advance was used to encounter the deep zones of weathering.

Plate load and pressuremeter tests were run at various locations along both crown drifts. Bearing capacity, strength, and moduli were determined. Figure 4 shows the results along the Honolulu-bound tunnel. A steady increase in bearing capacity occurred with distance from the portal. It was possible to locate the start of each weathering state from the geological mapping and the testing results. The testing program results indicated that tunneling in saprolite (extremely to highly weathered basalt) would be limited to a few hundred feet in each tunnel. The results also showed that the problems encountered in the Wilson Tunnel were unlikely to occur if the contractor used good construction procedures.

Rock Cores Versus Exploratory Tunnel. Figure 2 shows how boring NH-5 was very misleading. Based on the core recovery and logs, the material was classified as slightly weathered to unweathered. The core had decreased weathering states with depth, but the states were less weathered than predicted by the surface materials in a large tunnel face exposure. The exploratory tunnel through the crown of the main tunnels showed that boring NH-5 was drilled through a dike with strong, less weathered rock properties than the surrounding materials. Completion of the exploratory tunnel clearly indicated a gradual and uniform weathering profile through the mountain (Parsons Brinckerhoff-Hirota Associates, 1990). This profile was much more weathered than the boring suggested. Direct comparison of borings NH-4 and NH-5 with the tunnel geology was possible, since the borings were located within the tunnel envelope

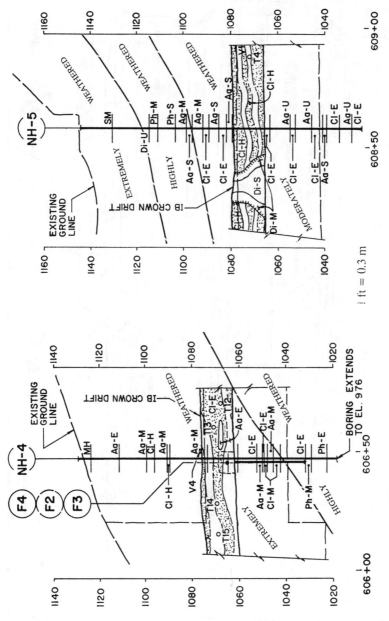

Figure 2. Schematic Showing the Information Obtained from Borings and Mapping at Two Locations in the Inbound Crown Drift.

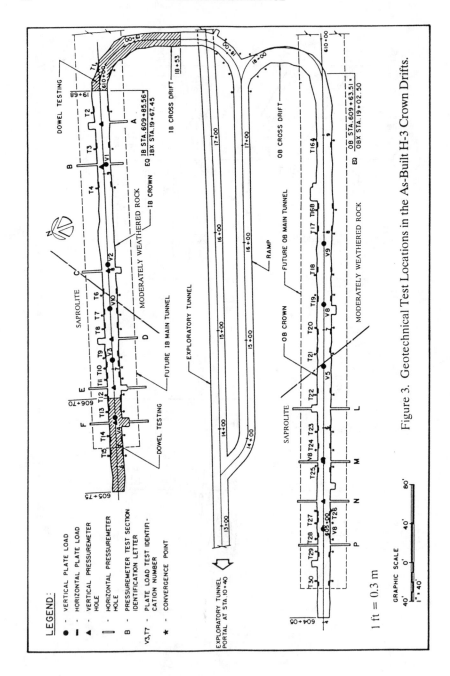

Figure 3. Geotechnical Test Locations in the As-Built H-3 Crown Drifts.

1 ft = 0.3 m

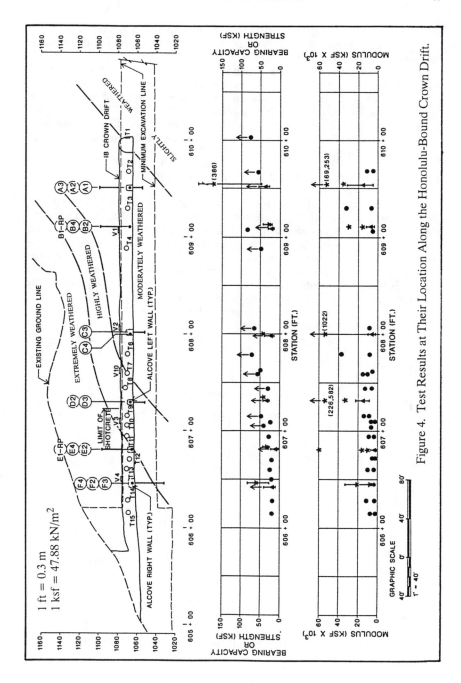

Figure 4. Test Results at Their Location Along the Honolulu-Bound Crown Drift.

and were intercepted during construction of the exploratory tunnel. These two borings serve as a reminder that borings often do not represent the materials in the immediate area where the boring is drilled.

The other lesson was that the drilling process may destroy the fabric of the rock. This means mechanical breaks are formed, and when not properly detected and logged, the rock is shown to have a lower RQD than is actually the case. For the H-3 project, the drilling process washed away the softer, weaker materials, leaving behind harder, less weathered corestones. This phenomenon is clearly shown in Figure 2, when comparing the log of Boring NH-4 with the geology mapped within the tunnel.

PROVO CANYON U.S ROUTE 189 TUNNEL PROJECT

The second highway project is the U.S. Route 189 in Provo Canyon, Utah. The project consists of widening a two-mile-long segment of U.S. 189 in a section between Wildwood and Vivian Park, Utah called the "Narrows," as shown Figure 5. The existing two-lane facility is being expanded to four lanes along the Provo River. The terrain consists of a series of steep rock ridges and cobble-filled talus valleys. Most of Provo Canyon is wide enough to accommodate four lanes with shoulders and median without crowding the adjacent river or creating excessively high rock cuts, fills, or retaining walls in this environmentally sensitive area. At the project site however, rock tunnels and high talus slope cuts are required for the expansion through the Narrows.

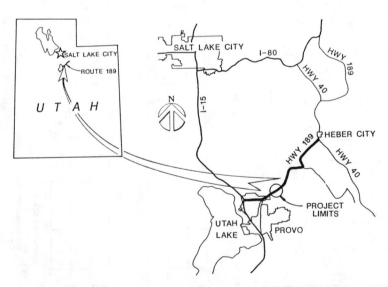

Figure 5. Location Map of the U.S. Route 189 Highway Project in Provo Canyon, Utah.

The tunnels will be driven parallel to the canyon wall. The canyon wall is fluted, with each flute filled with talus deposits. The 3-dimensional extent of the talus shoots is required to locate, design, and construct the tunnels and retaining walls. Geophysics, geologic mapping, and drilling were the methods used to define the extent and interface of talus and rock.

The project is located in the Wasatch Mountain Range. The terrain in this area is generally rugged, consisting of steep, stream-cut valleys, which terminate in narrow ridge crests. The alignment is located along the north side of Provo Canyon, which is generally oriented northeast to southwest. The Provo River occupies the bottom of Provo Canyon, flowing to the southwest.

The Wasatch Range in the area of the site is dominated by the Pennsylvanian age Oquirrh Formation. More recent alluvial and colluvial deposits occur in the canyon bottoms and as a veneer covering the bedrock along the steep canyon slopes. The Oquirrh Formation consists predominantly of limestone and of limey to quartzitic sandstone. Geologic units present along the alignment generally consist of surficial talus, alluvial, and colluvial deposits and the Oquirrh Formation bedrock (Chen-Northern, 1991).

Design Problem. The talus is present throughout the project alignment at the base of most large outcrops. It forms steep, generally active slopes, abundant on the west end of the project. The talus is made up of loose gravel and cobble and boulder size material. It becomes sandy and silty with depth. Talus slopes are steep and slope material tends to move downhill. The base of the slopes has been steepened by road cuts for the existing highway.

Colluvium covers a large portion of the site as well. It is not mapped in all areas where it occurs but only where significant accumulations are present. The colluvium varies greatly in composition from silty sand to gravel with cobbles and boulders. The amount of fines within the colluvium typically increases with depth.

Determination of where colluvium and talus makes contact with rock is a critical issue for the Provo Canyon Tunnel Project. Environmentalists are concerned about the impact a highway will have on the Canyon. By widening the existing highway, cuts and tunnels will be required. Knowledge of where rock begins will result in accurate definition of the extent of steep road cuts; minimizing the amount of excavation and limiting the impact. Determining where the talus and rock contact is located will provide a basis for reducing the amount of surface area of the retaining wall panels. Mapping, seismic surveys, and borings helped to define where the contact is located.

Geophysics, Mapping, and Exploration. Exploration of the site was composed of a three-step process consisting of geophysical testing, geologic mapping, and vertical and inclined core borings. The geotechnical exploration program was limited by the following constraints:

- Early information was required to confirm the feasibility of tunnel construction at the site.
- Environmental effects of drilling had to be avoided or reversed.
- The nesting period of golden eagles in the canyon began on January 20. No exploration activities were to disturb the eagles.

- Scheduling required that field work be carried out during the cold and snowy winter months at the floor of the canyon.

The geophysical testing, which consisted of seismic refraction techniques, was used to approximate the limits and characteristics of the talus and rock in a short amount of time. The testing results were then used to direct the drilling program. The geologic mapping was used to determine the areal extent of the geologic formations (fill, alluvium, colluvium, talus, and Oquirrh limestone) and the characteristics of the rock mass. The boring program was then designed to confirm the depth of talus, depth of bedrock weathering, joint filling characteristics, and location and characteristics of faulted and sheared zones.

The geophysics and mapping were done using conventional methods. The borings were done using lightweight wireline drill rigs flown in by helicopters onto prefabricated wooden platforms. The drill sites were restored, replanted and reseeded, irrigated, and cleaned of all spoil and contaminated soil. All noisy work was completed before the eagles landed.

Although it was feared that reflection off the ridge contours, near the V-shaped valleys would mask the overburden thickness, this did not prove to be the case. The borings in general confirmed the thickness of overburden indicated by the seismic refraction results (Figure 6). It was also feared that drilling through the talus would be impossible. This was not the case. The wireline drilling equipment was successfully used because the hole is cased by the drill string as it is drilled. With good bits and a cased hole, it was much easier than expected to determine the depth and characteristics of the talus as well as the underlying bedrock. Because the project costs were predominated by tunnel and open-cut excavation and support costs, it was imperative to define accurately the limits of each. The State of Utah had a small budget for the project and economic feasibility hinged on the results of these studies.

Design Solution. The new westbound lanes will traverse the Narrows with a series of tunnels through the rock ridges and retained cuts in the talus-filled valleys. The extent, cost, and visual impact of the retaining walls was a major and emotionally-charged issue during design. Minimizing the extent of the walls was most important, even though the tunnel engineering was of greater technical challenge. To minimize the size of retaining wall faces, a steep slope of 1H:14V was chosen. It was decided that even though vertical walls would result in the minimum face area, a slight batter was needed to reduce the visual impact.

Most of the cuts will be through the talus or overburden and into the underlying bedrock. While the geotechnical information provided a solid basis for designing the walls, their exact extent could not be accurately determined. Therefore, a retaining wall system was needed to achieve the necessary visual effects but which also could be adjusted in the field as actual geologic conditions were uncovered during construction. The selected solution includes cement grouting of the loose gravelly soils, shotcrete surface protection, dowel reinforcement, and cast-in-place or precast concrete facade with textured architectural coloring and treatment. Figure 7 shows a typical section of the solution. In addition to the geotechnical information used, CADD-generated cross-sections and profiles were invaluable in designing these walls and making ground checks in the field.

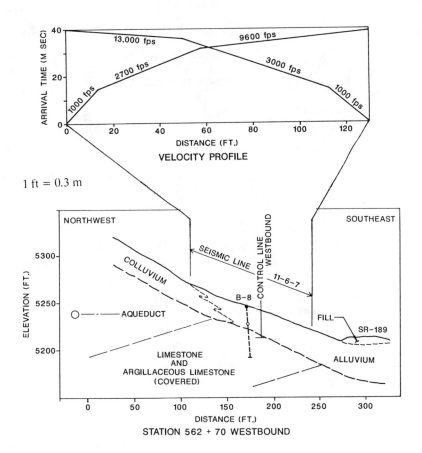

1 ft = 0.3 m

Figure 6. Schematic Showing the Information Obtained from Geophysics and Borings to Locate the Colluvium/Rock Interface.

GLENWOOD CANYON I-70 TUNNEL PROJECT

The third project is the Interstate I-70 Hanging Lake Tunnel Project in Glenwood Canyon, Colorado. Figure 8 shows the project location, which is east of Glenwood Springs. Twelve miles (19.3 km) of U.S. Route 6 is being converted into a four-lane section of Interstate 70. The project includes 5.7 lane-miles (9.2 lane-km) of viaduct, 1.6 miles (2.6 km) of tunnels, and 17.8 lane-miles (28.6 lane-km) of roadway included grade-separated sections. The two tunnel projects consist of the westbound Reverse Curve Tunnel, a 600-foot (183-meter) long, two-lane bore through a nose of rock, and the Hanging Lake Tunnels consisting of twin 40-foot (12.2-meter) diameter, 3,500-foot (1067-meter) long mined

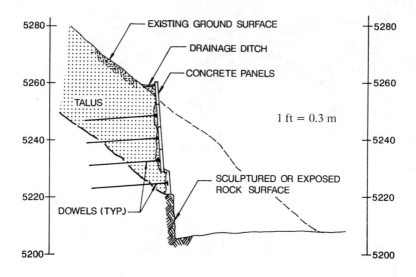

Figure 7. Proposed Retaining Wall System to Minimize the Impact of Highway Cuts in Provo Canyon.

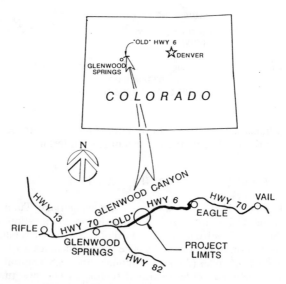

Figure 8. Location Map of the Interstate 70 Highway Project in Glenwood Canyon, Colorado.

tunnels, and a 500-foot (152-meter) long cut-and-cover tunnel, ventilation, and control facility called the Cinnamon Creek complex. The Reverse Curve Tunnel was completed in spring 1989 and the Hanging Lake Tunnels construction began in early 1990. The project site is located about 140 miles (225 km) west of Denver.

The geology within Glenwood Canyon consists of a complex sequence of pre-Cambrian igneous intrusive and metamorphic rocks that underlie the bedded Sawatch quartzites and other sedimentary rock formations that form the prominent cliffs of the canyon (Woodward-Clyde Consultants, 1985). The Hanging Lake tunnels were driven through massive to very blocky and seamy diorite.

Design Problem. The Hanging Lake Tunnels start at the Amphitheater portal, located on the east end, as the highway crosses the Colorado River. The tunnels extend through the massive rock and intersect and daylight at Cinnamon Creek. Cinnamon Creek, a tributary of the Colorado River, is lined with alluvium, colluvium, and talus deposits. The tunnels resume in rock beyond Cinnamon Creek. The tunnels continue to the western terminus known as the Shoshone Portal, where they daylight into a talus slope perched above an active railroad.

To handle the daylighting tunnels at Cinnamon Creek, a cut-and-cover section will be constructed. This section will join the two tunnel halves. Constructed as part of the cut-and-cover section will be the tunnels' ventilation facilities.

Since the tunnels intersected the creek, the critical design problem at the tunnel portals was determining where rock began. To define the thickness of the deposits, a 12-foot (3.7-meter) exploratory tunnel was constructed through the mountain.

Exploratory Tunnel and Mapping. Geologic exploration included surficial geologic mapping in 1978 and exploratory boreholes and laboratory testing in 1981. The exploration showed the presence of thick deposits of alluvium, colluvium, and talus at Cinnamon Creek and the Shoshone portal. The focus of exploratory boreholes was on the tunneling conditions, since most of the project cost related to the tunnels.

An exploratory tunnel was driven along the future eastbound tunnel alignment starting in 1984. The contractor constructed the Cinnamon Creek portals in talus by driving channel forepoles, then hand excavating. The portal excavations were stabilized with breast-boards and fully-lagged steel sets (Woodward-Clyde Consultants, 1985). Longitudinal sections of the Cinnamon Creek portals are shown in Figure 9. The exploratory tunnels showed that rock began 40 feet (12.2 meters) from the working portals.

Design Solution. The extent of the talus and alluvium deposits at the four portals was determined and documented. Most of the alluvium and talus was removed before construction began. However, some talus along the rock walls was left in place. The contractor was given the option of stabilizing the talus deposits and start mining the main tunnels or removing the talus deposits and then start the tunnel drives. The contractor chose to remove the talus material because the thickness of the deposits had been accurately determined during

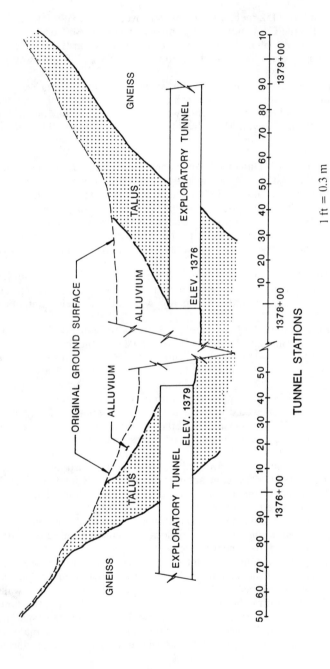

Figure 9. Schematic Showing the Information Obtained from the Exploratory Tunnel to Locate the Talus/Rock Interfaces at Cinnamon Creek.

construction of the exploratory tunnel. By knowing precisely where rock began, the contractor was able to make financial comparisons. In the final analysis, this information went a long way toward preventing construction problems and saving the owner money.

A second problem occurred, which involved the foundation conditions for the Cinnamon Creek complex. The complex is a special structure when considering the multiple sets of fans and motors used for ventilating the highway tunnels. In addition to the usual foundation loads of an industrial building, a significant dynamic component had to be considered from the fan vibrations particularly during start-up and shutdown. To avoid any potential problems because of differences in the talus and rock foundation characteristics, a minimum soil bedding underneath the building was required. The special soil layer was provided so the building could settle evenly. The depth of over-excavation of the rock was determined in the field.

DISPUTES REVIEW BOARDS AND DESIGN SUMMARY REPORTS

On these three tunnel projects, Disputes Review Boards and Design Summary Reports as outlined by the Underground Technology Research Council (1991) were used. Use of these items has allowed the owner and contractor to share the uncertainty and risk of defining where rock begins. The engineer is able to document where the soil/rock interface occurs based on exploration and experience. This documentation establishes a baseline, allowing the contractor to be compensated should different ground conditions be encountered.

CONCLUSIONS

Based on the experience of these three highway-tunnel projects, the following conclusions are made:

- Determination of where rock begins is a critical factor in design and construction and is best verified in the field once the actual geological conditions are exposed. This means the design must be flexible enough to handle actual field conditions.
- Designs should be flexible and well-conceived to handle problems that can arise in the field during construction.
- The design engineer should establish in the contract documents where he believes rock begins. These documents help to establish the baseline. If ground conditions are different and "rock" begins somewhere else, the contractor or owner can be compensated for the changed conditions fairly and without costly litigation.
- Borings alone are usually not effective in determining where rock begins. Drilling techniques can mask the in situ materials as well as underestimate lateral and vertical variability.
- Physical inspection and testing in situ is the best method of determining where rock begins.

- Use of Disputes Review Boards, Design Summary Reports, and Escrow Bid Documents can be very helpful in avoiding problems associated with determining where rock begins.

REFERENCES

1. L.W. Abramson and W.H. Hansmire, "Geotechnical Exploration for the H-3 Highway Trans-Koolau Tunnel," Proceedings of the Rapid Excavation and Tunneling Conference, Los Angeles, AIME, pp. 40-64, June 1989.
2. G.M. Boyce and L.W. Abramson, "Plate Load and Pressuremeter Testing in Saprolite," Geotechnical Special Publication No. 27, Volume 1, ASCE, pp. 52-63, 1991.
3. Chen-Northern, "Report of Geologic and Geotechnical Investigations, U.S. Highway 189, Provo Canyon Tunnels, Project F-019(30)," prepared for Parsons Brinckerhoff Quade and Douglas, March 1991.
4. D.U. Deere and D.W. Deere, "The Rock Quality Designation (RQD) Index in Practice," Rock Classification Systems for Engineering Purposes, ASTM STP-984, L. Kirkdaldie, Ed., ASTM, Philadelphia, pp. 91-101, 1988.
5. Parsons Brinckerhoff-Hirota Associates, "Halawa Approach and Tunnels Geology," Interstate Route H-3 Halawa Approach and Tunnels, FAIP No. I-H3-1(64), prepared for State of Hawaii Department of Transportation Highways Division, August 1990.
6. R.B. Peck, "Weathered-Rock Portion of the Wilson Tunnel, Honolulu," Soft Ground Tunneling, Failures and Displacements, D. Resendiz and M. P. Romo, Eds., Rotterdam, A. A. Balkema, pp. 13-22, 1981.
7. Woodward-Clyde Consultants, "Exploratory Tunnel, Geologic and Geotechnical Investigations," prepared for Department of Highways, State of Colorado, February 1985.
8. Underground Technology Research Council, "Avoiding and Resolving Disputes in Underground Construction, Successful Practices and Guidelines," Revised Edition, ASCE, 82p., 1991.

ACKNOWLEDGEMENTS

The authors would like to thank the Hawaii Department of Transportation, Utah Department of Transportation, Colorado Department of Highways, FHWA, and Parsons Brinckerhoff for their assistance and support of this paper. In particular, thanks go to Douglas Tanaka, Ed Keane, Ralph Trapani, William Hansmire, and Doug Slakey. Thanks to Boris Levitas for drafting the figures.

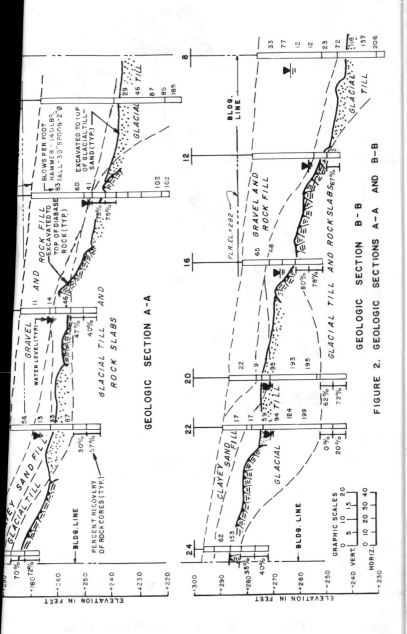

FIGURE 2. GEOLOGIC SECTIONS A-A AND B-B

GEOLOGIC SECTION A-A

GEOLOGIC SECTION B-B

FOUNDATION STUDIES
IN A FAULT ZONE AND A STEEP VALLEY SLOPE

By James L. Kaufman[1], P.E., Member ASCE and
Alfred H. Brand[2], P.E., Member ASCE

Abstract

This paper presents two case histories which
illustrate unique problems in defining the character of
rock for support of structures. The first case study
concerns a conventional building constructed on a site
where poor subsurface information existed at the time of
design. As a result, a conservative design was adopted.
As construction proceeded, the design was modified based
on the quality of the rock exposed. This type of
construction required full-time inspection by a competent
engineer who could make field decisions.

The second case study is for a project which was
recognized to be a difficult foundation problem. The
project was a bridge with large spans and heavy loads.
The site was difficult because it included the steep
slopes of a deep valley. The engineers had a free hand
in developing a foundation exploration program. The
initial exploration defined the general site characteristics. A second phase investigation explored the
individual pier and abutment foundation areas for the
proposed bridge.

FORT LEE APARTMENT HOUSE

Background

In the early 1970's, Mueser Rutledge Consulting
Engineers (MRCE) was asked to study the subsurface

[1]Senior Associate, Mueser Rutledge Consulting Engineers,
708 Third Avenue, New York, New York 10017.
[2]Partner, Mueser Rutledge Consulting Engineers, 708 Third
Avenue, New York, New York 10017.

conditions for a proposed apartment house in Fort Lee, New Jersey. The study included the examination of soil samples and rock core and a site visit. Historical and geological data concerning the site area were reviewed, but the borings were not inspected.

Subsurface Conditions

Geologically the site is in an area underlain by Palisades diabase. The diabase is an igneous flow onto the Triassic sandstone and shale underlying the Palisades area and the area west of the Palisades ridge. The diabase is locally known as trap rock and is an extremely hard rock used as a local source of aggregate. Numerous faults occur within the diabase, one of which was known to exist east of the project site. Glacial till covers the diabase ridge. The depth of till is variable and is controlled by local variations in the bedrock surface and the area drainage patterns.

The site is at the northern end of a surface depression draining to the south and then west to some meadows. Bedrock outcrops to the north and southwest of the property were noted during the site reconnaissance. Figure 1 is a site plan showing site topography and boring locations. The original ground surface sloped towards the south from Elev. 290 ft. (88.5 m) to Elev. 250 ft. (76.3 m) with reference to USGS, Mean Sea Level Datum.

An interpretation of the soils encountered in the borings is shown in the form of schematic Geologic Sections A-A and B-B on Figure 2. These sections are shown in plan on Figure 1. The stratification representing the conditions assumed between borings is shown as dashed lines. Information for each boring includes Standard Penetration Test (SPT) N-values in blows per foot, depth and recovery of cored rock, and water levels.

A classification of the soil samples indicated three major soil strata exist at the site:

Fill - An extensive depth of gravel and rock fill was found over the eastern side of the building. The boring contractor noted that the material required drilling to penetrate. Clayey sand fill was found underlying the gravel and rock fill at the western side of the building. Evidently, the site had been used as a disposal area for rock spoil

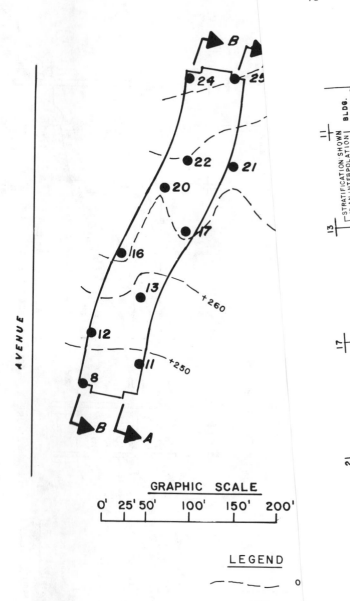

GRAPHIC SCALE

0' 25' 50'   100'   150' 200'

LEGEND

12   B(
      NU

FIGURE I. BORING LOCATION P

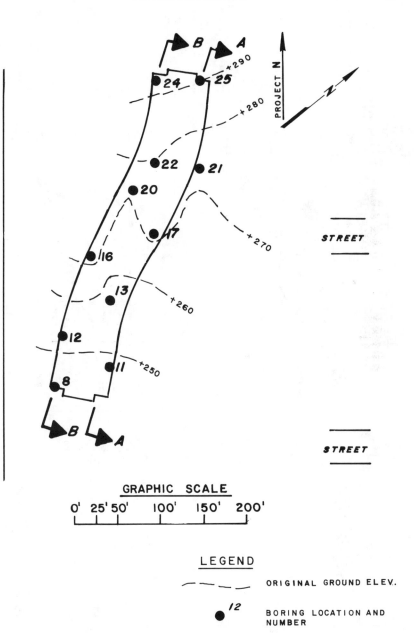

GRAPHIC SCALE

0' 25' 50'   100'   150' 200'

LEGEND

– – – – – ORIGINAL GROUND ELEV.

● 12 BORING LOCATION AND
       NUMBER

FIGURE I. BORING LOCATION PLAN

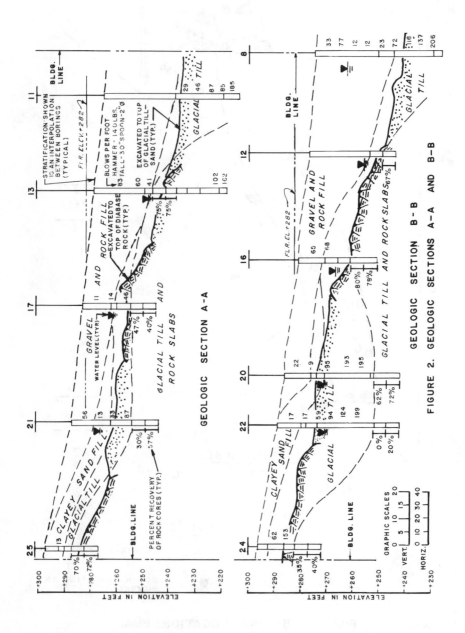

FIGURE 2. GEOLOGIC SECTIONS A-A AND B-B

removed during the excavation for bridge abutments
for the George Washington Bridge.

Glacial Till - Compact silty sands with some gravel
and trace clay. These soils were found beneath the
fill and are typical of the glacial debris found
overlying the bedrock. SPT's indicated N-values of
over 30 blows per foot.

Glacial Till and Rock Slabs - Extensive lengths of
rock were cored below the glacial till stratum. The
rock was a diabase similar to the bedrock in the
area. It was apparent that some of the rock cored
beneath the building's footprint was not bedrock as
glacial till was found beneath the cored rock.
Comparison of this rock with the outcrops on either
side of the site indicated differences in granular
structure and mineral composition. Therefore, it
is likely that the site was underlain by a fault
zone within the Palisades diabase.

Ground water levels recorded in the borings ranged
from Elev. 256 ft. (78 m) to Elev. 271 ft. (83.5 m),
generally near the surface of the glacial till
statum except at the south side of the site where
the water level was as much as 15 feet (4.5 m) above
the glacial till.

Foundation Design

The lowest floor was at Elev. 282 ft. (86 m), as
much as 25 feet (7.6 m) above the glacial till bearing
stratum. Based on the difficulties penetrating the fill
in the boring program, piles probably could not be driven
through the fill. Therefore, a recommendation was made
to excavate the rock fill and support column loads on
spread foundations in the underlying glacial till stratum
at an allowable bearing pressure of six tsf (600 kPa),
with a structural slab to support the lowest floor slab.

Foundation Construction

The subgrades for footings were inspected by MRCE.
Elevation and type of material encountered at the base
of the fill at footing locations were recorded on an as-
built plan. The solid stratification lines shown on
Geologic Sections A-A and B-B (Figure 2) reflect the as-
built plan. A rock line symbol is used to show where
apparent sound diabase was found at subgrade. The dotted

symbol is used to denote where sandy glacial till was
found at subgrade.

Conditions disclosed by interpolation between the
borings, shown as dashed lines on Figure 2, were
significantly different from conditions encountered
during construction. At the north side of the site,
apparently intact rock was found from five feet (1.5 m)
above to three feet (1 meter) below levels reported in
the borings. The rock surface sloped gently towards the
south from the north building line to about 90 feet
(27 m) south of the building line where the rock surface
plunged steeply toward the south. The rock exposed in
the northern quarter of the building was massive and
sound. Therefore, a redesign was recommended for the
footings on rock to a maximum bearing value of 20 tsf
(2,000 kPa). Due to the possibility of soil beneath
exposed rock, it was unadvisable to use higher bearing
values. The excavation contractor found that the sloping
rock could not be benched without blasting. It was
agreed to allow rock surface slopes as steep as IV:3H
provided that the rock surface was thoroughly cleared of
any surface weathering that would limit friction between
rock and poured concrete. Due to elevation changes of
several feet within some footings, the contractor formed
mass concrete leveling pads. The concrete footings were
then formed and poured at a later date.

The middle portion of the building's footprint was
founded on glacial till. This sandy till had no clay
binder and was found to be quite susceptible to subgrade
disturbance, due to the upward flow of groundwater. The
excavation contractor worked the foundation excavation
from north to south and used sumps and shallow collection
trenches to attempt to limit groundwater flow downward
to the south. Wellpoints were deemed inappropriate due
to the highly variable bedrock surface. The contractor
could maintain a stable subgrade below the water table
only by excavation to subgrade immediately before pouring
a six-inch (150 mm) thick working mat of lean concrete.
A backhoe would excavate, without teeth to subgrade.
Laborers then placed a form around the subgrade and
immediately started excavating a shallow trench on the
outside of the form. Any loosened soils at subgrade were
removed as they appeared and the working mat was poured
as soon as the inspector approved the subgrade. The
actual footing was formed and poured at a later date.

Potential differential settlements were a concern
where footings supported on rock were adjacent to
footings supported on soil. Therefore, the contractor

was instructed to create a transition zone. All footings on rock, adjacent to footings on soil were required to have a six-inch thick (150 mm) tamped sand cushion placed over the concrete leveling course, prior to footing placement. These footings were restricted to a maximum bearing value of six tsf (600 kPa). The sand cushion was retained by a concrete curb poured with the concrete leveling course.

A rock nob was encountered in the southern half of the building. While rock levels generally agreed with the borings, the rock surface was quite a bit more irregular than indicated in the borings. The exposed diabase was massive and sound. The southern quarter of the building was founded on glacial till. Transition zones with sand cushions were employed over rock wherever there was an adjacent footing founded on glacial till.

Substantial overexcavation to rock occurred in the southern portion of the structure as a result of the contractor's inability to obtain a stable subgrade in the sandy glacial till.

Conclusions

Based on the linear stratification of soils encountered in the borings, it was anticipated that the structure would be founded at the surface of the glacial till stratum at an allowable bearing pressure of six tsf (600 kPa). As a result of a highly variable rock surface, and the overexcavation that occurred in the portions of the southern section of the building, approximately one-half of the footings were founded on rock that probably represents thick rock slabs within a fault zone. However, by taking advantage of the rock's ability to distribute loads, even if underlain by soil, the rock bearing value was increased to 20 tsf (2,000 kPa).

GLADE CREEK BRIDGE

Background

In the early 1980's, MRCE performed a two-phase foundation investigation for the Glade Creek Bridge on Interstate I-64, approximately nine miles east of Beckley, West Virginia. Work was performed under a subcontract agreement between MRCE and Howard Needles Tammen & Bergendoff (HNTB). The owner is the West Virginia Department of Transportation, Division of Highways.

The Glade Creek Bridge site is a steep-sided V-shaped valley about 2,500 feet (763 m) wide and 700 feet (214m) deep. The creek's bottom is at Elev. 1904 ft. (580.7 m) (USGS Datum) and the valley top is at Elev. 2630 ft. (802.2 m). Valley side slopes average 1V:1.5H with local vertical rock exposures. An initial investigation was conducted to determine the general character of the bedrock profile along the bridge alignment. The final investigation was to evaluate in detail the pier and abutment foundation areas for the prestressed concrete box girder bridge alternative. Concurrently, a continuous steel deck truss bridge alternative was being designed by others. The steel bridge was eventually constructed in the late 1980's.

Area Geology

The site is on the Raleigh-Fayette Plateau, a part of the Appalachian Plateau. The topography is characterized by rugged low mountainous terrain. Due to the relatively flat-lying rock strata, a well defined dendritic (branching) drainage system has developed. The rock formations are Pennsylvanian to Mississippian age. These consist of sandstone, shale, conglomeratic sandstone and coal seams. The regional strike of bedding planes is North $40°$ East, with the regional dip of the beds varying from $2°$ to $5°$ towards the west.

In sequence with depth, valley stratigraphy commences with an approximate 550-foot (168 m) interval of the lower Pennsylvania system divided into two formations: the New River, typically 50 ft. (15 m) to 150 ft. (46 m) thick, overlies the Pocahontas which is on the order of 400 ft. (122 m) thick. The upper Mississippian system is represented in the lower portion of the valley by the Bluestone formation. Generalized Stratum Descriptions for the site vicinity are shown on Table 1.

Valley Stress Relief

Understanding the mechanism and result of valley stress relief for the interbedded sandstones and shales is critical in evaluating the stability of near-surface rock zones. Stream erosion carves into the valley profile and removes lateral and vertical support from the valley sides. Stress relief accompanying the removal of lateral support results in tension fractures and bedding plane shear zones in rocks adjacent to the valley walls. The near-surface fractured layer or zone of extension exhibits the following typical characteristics for sand-

TABLE 1   **GENERAL STRATUM DESCRIPTIONS**

SYSTEM	FORMATION	STRATUM	DESCRIPTION
PENNSYLVANIAN	NEW RIVER	SH 1	GRAY SANDY TO SILTY SHALE WITH SANDSTONE BEDS.
		SS 2	GRAY SANDSTONE, SOME SHALE STREAKS.
	POCAHONTAS	SH 2	GRAY SANDY SHALE LAMINATED WITH SANDSTONE.
		SS 3	GRAY TO RED SANDSTONE INTER-BEDDED WITH SAND TO SILTY SHALE.
		SH 3	GRAY TO GREEN SANDY TO SILTY SHALE.
		SS 4	GRAY BROWN SANDSTONE AND CONGLOMERATIC SANDSTONE.
MISSISSIPIAN	BLUESTONE	SH 4	GRAY TO RED SILTY TO SANDY SHALE.
		SS 5	GRAY BROWN SANDSTONE.
		SH 5	GRAY AND RED SILTY SHALE, SOME SILTSTONE BEDS.

stones and shales, depending on the strength of the beds:

1.   Weaker (shale) beds tend to develop diagonal to curved shear joints and commonly develop clay-filled seams at contacts with strong beds.

2.   Stronger (sandstone) beds develop vertical to sub-vertical tension joints which do not extend across weaker beds or bedding contacts.

3.   Valley stress relief accelerates valley wall weathering as the rock mass loosens through the development of joints and fractures and groundwater flows through the stress-relief fractures.   Differential weathering and accompanying rock falls are common where strata are exposed and weaker beds erode, leaving overhanging cliffs of stronger rock layers.

Phase 1   Field Investigation

Boring Program -   The program consisted of eight, 250-feet (76.2 m) deep borings and one 150-feet (45.8 m) deep boring spaced to allow the rock profile to be intersected for the valley's full height. Approximately 50 feet (15 m) of core overlap between adjacent borings aided to establish continuity and the dip of individual rock beds.

Concurrent with the boring work, HNTB retained Weston Geophysical Corporation to perform a seismic investigation along the bridge alignment. Surface refraction lines parallel and perpendicular to the alignment were performed to determine the character and variation of near-surface deposits. Downhole velocity surveys were performed in selected boreholes to establish stress-strain parameters of the various rock strata.

For purposes of brevity, discussion is limited to the east side of the valley. As-drilled borings are shown in plan on Figure 3.

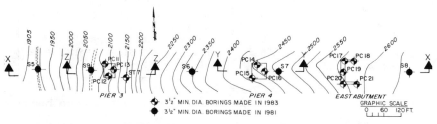

FIGURE 3 BORING LOCATION PLAN, EAST SIDE OF VALLEY

Borings were advanced with four-inch (102 mm) casing through overburden soils which were sampled with a conventional two-inch (51 mm) O.D. split-spoon. Upon encountering split-spoon refusal, the casing was seated into rock. The boring was then advanced to completion using a double-tube NQ wireline core barrel recovering 1-7/8-inch (48 mm) diameter core. Individual core runs were usually limited to five feet (1.5 m).

Seismic Surveys - Downhole velocity surveys were successfully performed in Borings Nos. S-2 and S-7. Surface refraction surveys were performed on the upper half of the valley sides. These seismic lines included surveys along and perpendicular to the bridge alignment.

Phase I   General Subsurface Conditions - Geologic Section X-X, as shown on Figure 4, is a natural scale

section of the east side of the valley. The alternating
strata of sandstones and shales are separated into nine
general strata as described on Table 1. In addition,
based on the seismic work, three zones roughly parallel
to valley slopes were distinguished:

Overburden. The borings encountered a three to 13-
foot (1 to 4 m) thick layer of overburden materials. The
overburden is a colluvium and its characteristics depend
on the nature of the parent rock. When derived chiefly
from shale, the soil is a stiff to hard brown to gray
silty clay to clayey silt with some shale fragments.
When derived from sandstone, the overburden is a medium
compact fine to coarse sand, some sandstone fragments,
trace organics and roots; to brown to red-brown fine
sandy silt. Overburden materials were sampled by driving
a standard split-spoon. In many borings a roller bit was
required to drill through weathered and broken rock
fragments on top of the coreable bedrock surface. The
overburden zone was detected in the geophysical
investigations performed on the upper half of the valley
slopes. The layer is defined as the zone with a
compression wave (P-Wave) velocity between 1,200 feet per
second (366 mps) and 2,400 fps (732 mps). The thickness
of this zone varies from approximately three to 18 feet
(1 to 5.5 m) on the east to west valley slopes.

Broken to Fractured Rock. This near-surface zone
was identified in each of the borings as a slightly
weathered, fractured to broken rock zone with typically
more than four joints per foot. On both sides of the
valley this layer ranged from 10 to 40 feet (3 to 12 m)
thick. A continuous profile along the upper half of the
valley was identified from surface refraction surveys
performed by Weston Geophysical Corp. This zone was
defined as materials with P-Wave velocities of 4,500 to
6,500 fps (1,370 to 1,980 mps). On the east side of the
valley the layer was as much as 50 feet (15 m) thick.

The broken to fractured rock zone represents rock
which has been the most altered by stress relief
mechanisms. The stress relief fracturing may have
resulted in actual separation of rock blocks from the
more competent underlying bedrock. In this zone
extensive vertical jointing within the sandstone and
cross-jointing within the shale were anticipated. This
zone may behave more like a compact blocky coarse-grained
soil rather than bedrock.

Competent Bedrock. From the seismic survey this
zone was defined as rock having P-Wave velocities of

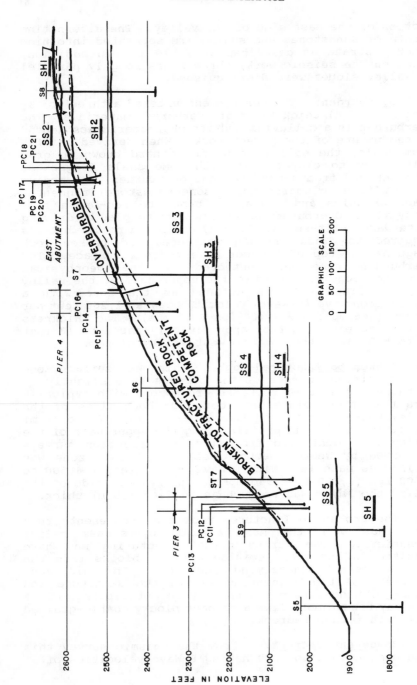

FIGURE 4. GEOLOGIC SECTION X-X

8,800 to 11,000 fps (2,680 to 3,360 mps). On the basis of the surface refraction and downhole velocity surveys, intact rock below the broken to fractured surface zones is generally quite competent with the shale and sandstone having similar compression wave characteristics.

Phase II   Field Investigation

The final exploration program was designed to evaluate specific foundation areas. Typically three borings, 80 to 100 feet (24.4 to 30.5 m) deep, were drilled at each pier. Five borings, 20 to 70 feet (6.1 to 21.4 m) deep, were drilled at the east abutment. The seismic program consisted of cross-hole surveys at both piers on the valley's east slope and seismic refraction lines at each pier and abutment. Goodman Jack tests were performed in boreholes at each pier location.

The boring depth criteria was to extend pier and abutment borings 50 feet (15.3 m) and 20 feet (6.1 m), respectively, into competent bedrock. In an attempt to determine vertical joint spacing, one boring at each pier and abutment was drilled at an angle of 15 degrees from vertical into the valley slope in a plane roughly at right angles to the surface contours. The cross-hole seismic program required a signal hole at a minimum distance of 50 feet (15.3 m) from the array of receiving holes. At Pier 4, provisions were made to clean out a boring from the earlier investigation for use as the signal hole. At Pier 3, an additional boring upslope of the pier was added for use as a signal hole.

Through overburden, the borings specified for cross-hole studies were advanced with four-inch (102 mm) casing and the remaining borings were advanced using hollow stem augers with conventional two-inch (51 mm) O.D. split-spoon sampling. Upon encountering split-spoon refusal, the casing or auger was seated in rock. The boring was then advanced to completion using a double-tube NX core barrel recovering 2-1/8-inch (54 mm) diameter core. Individual core runs were usually limited to five feet (1.5 m). Occasionally, the runs were up to 10 feet (3 m).

Phase II   General Subsurface Conditions

In the Phase II investigation, relatively closely-spaced borings were drilled at piers and abutments to further define the limits of the rock zones. In recognition of the fact that refraction techniques cannot readily disclose layers of soft or broken rock beneath

more competent rock layers, Phase II also included
crosshole surveys at Piers 3 and 4 to aid in defining the
variation for the rock quality profile.   The rock
structure zones at all piers and abutments were defined
by correlating the results of cross-hole studies with
actual core descriptions.

The as-drilled locations of the east valley pier and
abutment borings as shown in plan are on Figure 3.
Figures 5 and 6 show natural scale geologic sections at
Piers 3 and 4 respectively. The alternating sequence of
sandstone   and   shale   are   shown   as   solid   line
stratification.   Based on a detailed evaluation of the
physical rock core and seismic studies, the rock was
divided into a series of rock structure zones based on
rock competency as summarized on Table 2.   These zones
are shown as dashed lines on Figures 5 and 6.   The
average rock core recoveries and Rock Quality Designation
values for these zones are noted on the borings.

In developing the rock structure zones, a relatively
thick transition zone was noted between broken to highly
fractured   rock   and   the   competent   bedrock.   This
transitional zone is clearly identified in cross-hole
velocity profiles at Piers Nos. 3 and 4.

At Pier 3, below a thin mantle of overburden soils
(Zone I), the upper 35 feet (10.7 m) consists of a broken
to fractured rock (Zone II) with P-wave velocities that
vary from 5,600 to 6,500 fps (1,710 to 1,980 mps).
Immediately below, there is a transitional layer (Zone
III) of slightly weathered rock with some fractured zones
between depths of 40 and 60 feet (12.2 and 18.3m) with
P-wave velocities of 6,500 to 7,700 fps (1,980 to 2,350
mps).   Below a depth of 70 feet (21.4 m), in competent
bedrock (Zone IV), the compression wave velocities are
8,800 to 10,700 fps (2,680 to 3,260 mps).   Core
recoveries within Zone II vary from 34 to 98 percent.
The low core recoveries signify extensively weathered
material that washed out during the coring process. The
RQD values vary widely from 0 to 72 percent, averaging
15 percent.   Core recoveries within Zone III were
generally greater than 95 percent while RQD values ranged
from 0 to 80 percent with average of 50 percent. The low
RQD values signify highly fractured zones within this
generally competent rock.   Core recoveries within Zone
IV were generally greater than 95 percent and the RQD
values averaged 80 percent.   Limited fractured zones
generally less than one foot (0.3 meters) thick were
encountered in the competent rock mass.

A similar sequence of rock structure zones were developed for Pier 4 based on a comparison of the crosshole seismic survey and the physical rock core. At Pier 4, Zone I is 7 to 10 feet (2 to 3 m) thick. The Zone II surface is 14 to 35 feet (4.3 to 10.7 m) below ground surface. Core recovery in Zone II ranged from 25 to 98 percent with RQD values of 0 to 40 percent, averaging less than 10 percent. The transitional Zone III layer has a thickness of 23 to 35 feet. Core recoveries are generally very good with RQD values of 0 to 62 percent averaging 25 percent. The underlying Zone IV is found at 37 to 70 feet (11.3 to 21.4 m) feet below ground surface with generally excellent core recoveries and RQD values averaging approximately 55 percent.

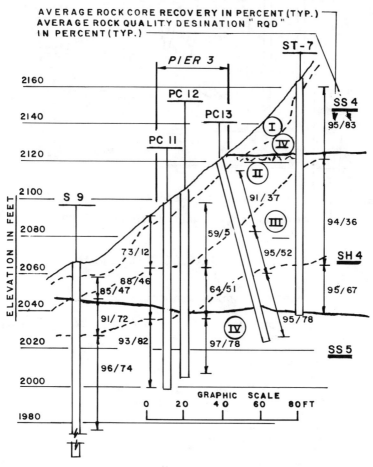

FIGURE 5. GEOLOGIC SECTION Z-Z

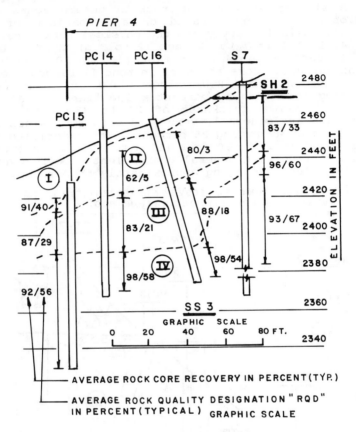

AVERAGE ROCK CORE RECOVERY IN PERCENT (TYP.)

AVERAGE ROCK QUALITY DESIGNATION "RQD"
IN PERCENT (TYPICAL) GRAPHIC SCALE

FIGURE 6. GEOLOGIC SECTION Y-Y

TABLE 2   ROCK STRUCTURE ZONES

ZONE	CLASSIFICATION	P-WAVE VELOCITY (FEET/SEC.)
I	OVERBURDEN SOILS	1,200 TO 2,400
II	BROKEN TO FRACTURED ROCK	4,500 TO 6,500
III	SLIGHTLY WEATHERED ROCK, SOME FRACTURED ZONES	6,500 TO 7,700
IV	COMPETENT ROCK, OCCASIONAL FRACTURED ZONES	8,800 TO 11,000

Results of Field and Laboratory Testing Programs

Separate field and laboratory testing programs were performed to evaluate qualitatively the strength and deformability characteristics of the rock mass. The field testing program included Goodman Jack testing, seismic refraction surveys and cross-hole seismic surveys. Laboratory testing was limited to unconfined compression tests of intact rock specimens.

The average Young's Modulus (E) determined from the Goodman Jack test data, generally increases with rock quality as described by the RQD. Zone II (Broken to Highly Fractured Rock) with an average RQD of 15%, displayed an average E of 0.5×10^6 psi (3×10^6 kPa). Similarly, Zone III (Slightly Weathered Rock, Some Fractured Zones) with an average RQD of 30% had on average E of 1×10^6 psi (7×10^6 kPa). Zone IV (Competent Bedrock, Occasional Fractured Zones) with an average RQD of 70% had an average E of 1.5×10^6 (10×10^6 kPa). The Young's Modulus determined from the cross-hole seismic surveys tended to be in the range of 1×10^6 psi (7×10^6 kPa) to 2×10^6 (14×10^6 kPa) and did not correlate well with the RQD. The cross-hole seismic results generalize elastic properties between the borings used in the test array and tend to mask local discontinuities as reflected by the RQD.

Zone III was predominantly sandstone and sandy shale. Sandstone beds found in Zone III were found to be of better quality than shale beds. Sandstone had average RQD values of 67% and 44% within the transition zone at Piers 3 and 4 respectively, while the shale had corresponding values of 44% and 24%. Zone IV rock, identified as competent bedrock and characterized by high RQD, was predominantly sandstone. The Goodman Jack and cross-hole seismic tests typically yielded higher values of elastic modulus for the sandstone as compared with the shale. Compressive strengths of intact rock specimens averaged 8,800 psi (61,000 kPa) and 21,000 psi (145,000 kPa) for the shale and sandstone respectively and clearly characterized the shale as the weaker member within the rock formation.

Conclusions and Foundation Recommendations

1. The subsurface investigation program delineated the valley slope stratification as well as the stratification at individual piers and abutments.

2. The investigation also defined the qualitative

rock structure zones at each pier and abutment necessary to determine bearing levels for the proposed foundations.

3.   The existing valley side slopes average 25° to 40° and appeared stable. The slopes are underlain by thin mantle of overburden over a zone of broken to highly fractured rock which varies in thickness from 14 to 41 feet (4.3 to 12.5 m) below the valley surface. Beneath this zone is a transition into competent bedrock with the transition zone typically 15 to 35 feet (4.6 to 10.7 m) in thickness over the fully competent bedrock. Some highly fractured and broken zones were found within the transition zone and occasional highly fractured intervals were encountered within the competent bedrock.

4.   Sandstone was generally found to be less fractured and of higher strength than the shale throughout the rock profile. It was concluded that the pier and abutment foundations should extend through the broken and fractured rock zone for support in the underlying transition zone at a maximum bearing intensity of 10 tsf (1,000 kPa) for predominantly shale and 15 tsf (1,500 kPa) for predominantly sandstone subgrades. The transition zone which is typically 15 to 35 feet (4.6 to 10.7m) in thickness was to be grouted for its full depth with a particulate grout to knit together the broken and fractured zones encountered in the layer.

5.   As an alternative to spread foundations, rock socketed caissons could be extended into the transition and competent rock zones. Allowable bond stress on the caissons was established at 80 psi (550 kPa) and 100 psi (690 kPa) in the transition zone and competent bedrock, respectively. No bond stress was allowed within the highly broken and fractured zone. Recommended end bearing values for the shale and sandstone were 10 and 15 tsf (1,000 and 1,500 kPa), respectively with an allowable increase of 10 percent per foot of embedment below the surface of the base of the broken rock zone up to double the minimum allowable values. The transitional zone was to be grouted for its full depth similar to the spread bearing alternative to knit together the broken and fractured zones.

Soil-Rock Transition Zone:
Uncertainties For Design and Construction

R. E. Smith[1] M. ASCE, M. A. Gabr[2] AM. ASCE,
and J. R. Kula[3] M. ASCE

Abstract

This paper addresses the problem of establishing definitions of the soil-rock boundary with regard to different engineering applications in the Piedmont geologic provinces. Three engineering situations illustrating criteria adopted for definition of soil-rock interface are presented. The present engineering state-of-the-knowledge to form such definitions in the Piedmont weathered profiles is inadequate. The construction industry would be best served by a clear delegation of risks and rewards of predicting excavatability to the contractor.

Introduction

In locations where geologic discontinuities have resulted in relatively soft soils overlying massive hard rock, the geometry of the soil-rock boundary can be reasonably defined with existing subsurface exploratory techniques. In areas of weathered and decomposed rock profiles, such as that of the Piedmont Physiographic province of the southeastern United States, definition of the soil-rock boundary is a recurring challenge for engineers and contractors. In this situation, the subsurface conditions typically consist of surface soils derived from extensive weathering of the parent rock. With depth, the soils grade into less-weathered material and more evidence of the parent rock features are retained. At some depth, no signs of weathering within the rock mass can virtually be detected.

In these types of transitional subsurface profiles, construction disputes over the issue of what is "rock" are

[1] Senior Principal, Woodward-Clyde Consultants, Maryland
[2] Project Engineer, Woodward-Clyde Consultants, Maryland
[3] Associate, Woodward-Clyde Consultant, Maryland
904 Wind River Lane, Gaithersburg, Maryland 20878

common between engineers and contractors. A primary reason for these disputes is the different functional definitions that engineers and contractors have for the word "rock".

For the engineer, the rock of interest is the material that will be left in place when construction is completed. It is important because its strength, compressibility and/or permeability affect the performance of the constructed facility. For the contractor, the rock of interest is that portion that must be removed during construction of the facility. Its importance is related to the methods, costs and time required to excavate, process and/or dispose of it. Hence, the word rock used in a geotechnical report may convey one meaning to the engineer and another to the contractor.

For excavation to proceed, rock must be broken into pieces that are small enough to be handled by excavation equipment. Therefore, excavatability of a rock mass is measured by the type of force systems that must be used to break the rock mass into appropriately small pieces in a reasonable amount of time. It is believed that excavatability is a function of such factors as: the unconfined compression of the rock mass, the shear strength of the rock discontinuities, the frequency of rock discontinuities, and the orientation of rock discontinuities relative to the excavation free face.

Relatively little research has been done to develop subsurface investigation techniques to aid in predicting excavatability, i.e., breakability. Investigation techniques are specially needed for materials that are transitional between easily excavatable, adhesionless soils and massive, hard rock that is without weakened discontinuities and requires patterned blasting to break for removal. Research to quantify the excavation characteristics, i.e. breaking or disaggregating, of various types and sizes of excavating equipment is needed. Work performed by Caterpiller company, as for example presented by Krukowski (1990), represents one of few significant efforts in this area.

In general, cost-effective excavation, whether for mass grading or drilled shaft installation, is dependent on two factors:

1. The cost of the action (blasting, ripping, etc.) required to break the in-place material into sizes that can be effectively removed.

2.   The cost of time required to do the breaking.
Excavation cost will be different depending on whether the
rock   is   slowly   ground   or   quickly   blasted   into   the
appropriate size for effective removal.

This paper, in view of "how and who?", investigates
definitions   of   the   soil-rock   boundary   for   different
engineering applications in the Piedmont geologic province.
Three general situations describing the identification of
the soil-rock boundary with case histories data are
presented. Various definitions of the soil-rock boundary
are discussed based on predictions using Standard
Penetration Test (SPT) N-values, Rock Quality Designation
(RQD), and seismic velocity.

Soil-Rock Interface: Previous Findings

Quantitative definition of the soil-rock interface
have been addressed in the literature. Coates (1970)
recommended that the Rock Quality Designation (RQD) value
could be used to estimate excavation depth before reaching
sound rock. RQD values smaller than 25% designate very poor
rock quality that could be classified as soil for
engineering   purposes.   Peck   (1976)   stated   that   the
distinction between rock-like and soil-like material in
transition zones is usually unpredictable. Figure 1, as
presented by Deere and Patton (1971), illustrates this
point for metamorphic and intrusive igneous rocks.

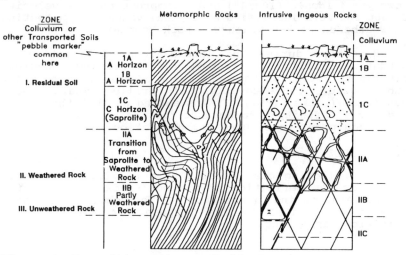

Figure 1. Transition Between Residual Soil and Unweathered
Rock (From Deere and Patton, 1972)

Peck also stated that while certain shales are geologically classified as rock, they would be considered heavily overconsolidated clays from the engineering point of view.

Obermeir (1979) presented typical compression wave velocities for estimating type of material in the Piedmont weathered profile. A range of 1700 to 2300 feet per second (fps) was suggested for the weathered rock zone and 3300 to 13,000 fps for the unweathered rock zone. Sowers and Richardson (1983) presented an idealization of a typical subsurface profile for the Piedmont area. The profile consisted of residual soil, transition material, weathered and jointed rock underlain by unweathered rock. The authors also presented typical engineering characteristics of layers identified in the profile. Partly weathered rock was identified as having a typical void ratio of 10% to 50%, and a permeability of 10^{-1} to 10^{-5} cm/sec .

White and Richardson (1987) recognized the difficulty of excavatability evaluation in transitional zones that have the characteristics of both soil and rock. The authors presented a summary of classification approaches, for Piedmont and Blue Ridge areas of the eastern United States, based on a survey of practices followed by several engineering consultants and contractors. The soil-weathered rock boundary was identified when an N-value range of 80 to 100 blows per foot (bpf) and a compression wave velocity of approximately 3500 fps were obtained. An N-value of 100 blows per 4 inches (100/4") and a compression wave velocity of 6000 fps were identified as typical values for detecting the weathered rock-unweathered rock boundary.

Recently, during the International Association of Foundation Drilling Contractors (ADSC) meeting, possible criteria to establish the definition of soil-rock boundary with regards to excavation of drilled shafts were identified (Litke, 1991). It was agreed on that there is a need for a definition to account for the material as well as the equipment to be used for excavation.

Case Studies

The following case studies briefly discuss criteria chosen for the definition of the soil-rock interface for various engineering applications. The choice of these criteria was driven by the purpose for which such a definition was needed.

Situation 1: Excavation for Buildings

Two cases involving rock definition, or specifically lack thereof, provide some insight into correlation between measured field parameters and excavation behavior. The first case involved the interface between soil, defined as material excavatable using scrapers and dozers but no rippers, and harder materials that require ripping but no blasting. After the excavation contractor achieved mass grading excavations of up to 15 feet without rippers, he ceased work claiming that the contract specified rock prices for any excavations requiring ripping or blasting. During the dispute negotiation period, the design engineering firm for the project conducted about 20 seismic refraction profile lines within approximately seven acre area. Our firm was retained to conduct an additional 10 seismic lines to independently evaluate the exposed excavation surface.

The two sets of the seismic compression wave data were generally consistent for the exposed excavated surface down to 10 to 20 feet below the existing exposed ground surface. Basically, the seismic velocity ranged from 1200 to 2300 feet per second (fps) with the majority of all values in the range of 1500 to 2000 fps. Pre-excavation N-values for the soil at the "no-ripping" surface averaged approximately 60 blows per foot (bpf).

The seismic compression wave values, 1500 to 2000 fps, obtained for the exposed surface zone were lower than the 3000 to 3500 fps often cited in literature for the soil-weathered rock ripping boundary. However, data from survey results reported by White and Richardson (1987) indicated that one engineering respondent specified 2000 fps for the onset of the ripping interface. It is of interest to note that test pits dug with a John Deere JD690 backhoe, equipped with a one cubic yard bucket with digging teeth penetrated the "no-ripping" surface by depths of 8 to 16 feet.

The second case involved the interface between ripping and blasting and/or hoe-ramming of a deep basement excavation. The building, located in Washington, D.C., has four below-grade basements. An excavation on the order of 60 feet was required to achieve the lowest design grade.

The site lies within the eastern Piedmont geologic province. A generalized soil profile is presented in Figure 2. The field investigation consisted of Standard Penetration Testing and rock coring. Samples from the borings indicated that the upper 10 to 20 feet of the site

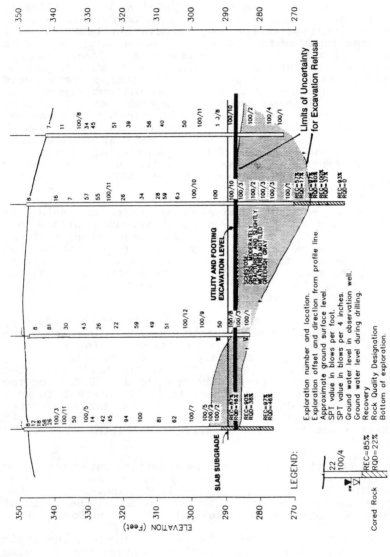

Figure 2. Typical Subsurface Profile: Excavation for Commercial Buildings

soils were either fill or alluvial sands and gravel. Below
this depth, the soils were typically residual consisting of
micaceous sandy silts and silty sand. The residual soil
became denser with depth with a gradual transition to
weathered rock. The N-value for the residual material
ranged from 25 blows per foot (bpf) to more than 100 bpf.

The rock core samples indicated that the rock is
generally heavily jointed, weathered, schistose gneiss.
Joint spacings were typically on the order of two to eight
inches. Table 1 present a summary of the field
investigation results.

Table 1. Summary of Field Investigation Results

Material	N-Value	Recovery	RQD
Residual Soil/Decomposed Rock	25-100$^+$ bpf	-	-
Rock Cores: Upper 5 Feet Rest of Core	- -	45%- 80% 15%-100%	6%-29% 15%-70%

A dispute occurred between the owner and the
contractor over the issue of how much rock the contractor
should have accounted for in his bid price. To help
evaluate "what the contractor should have known" from the
site data, our firm was retained to evaluate the subsurface
conditions and make an estimate of the amount of rock that
would likely require blasting and/or hoe-ramming based on
the pre-bid information. This evaluation was made without
the benefit of knowing the results of the completed
excavation.

Top of rock in terms of uncertainty margins was
defined as shown in Figure 2. The upper boundary was
defined as being rock-like material that would have a high
probability of being excavated by the Caterpiller 235
backhoe. The lower boundary was defined as being rock-like
material that would have a low probability of being removed
by the Caterpiller 235 backhoe. This boundary was estimated
to be at a level where the RQD generally exceeded 50%.
Actual excavation quantities did fall within these limits.
Approximately, 20,000 cubic yard (cy) of material required
ripping. Blasting was used to remove 1500 cy only in
isolated excavations for footings and utilities below the
subgrade line. The boundary between ripped and blasted
material was represented by an N value of approximately
100/2".

This example points out the significant uncertainties
involved in an engineering estimate of rock excavatability.
Three different contractors in the greater Washington area
defined the limit of rippability in decomposed rock profile
using a D-9 Dozer as shown in Table 3.

Table 3. Limit of rippability in decomposed rock profile
         using a D-9 Dozer: A Survey of Contactors

Contractor	N-Value
Contractor 1	100/3"
Contractor 2	100/3"-100/8"
Contractor 3	100/8"-100/10"

In this engineering situation, assessment of the soil-
rock extent was for excavation requirements. Therefore,
knowledge of material substance and formation was
important. It is not uncommon that a soft unjointed
material may offer more resistance to excavation than a
harder fractured material where ripping teeth can penetrate
into the fractures.

Situation 2: Excavation to Achieve Reduced Permeability

This situation involved a dam located in Maryland. The
dam is composed of a relatively impervious central core
that is protected by both upstream and downstream
transition zones. The core was extended down to a cut-off
trench under which a grout curtain, 30 to 60 feet deep, was
installed for seepage control.

The cut-off trench was excavated along the centerline
of the base of the dam. The width of the trench varied with
a minimum width of 20 feet and a maximum width of 40 feet.
In an attempt to minimize seepage beneath the dam and to
facilitate grouting of the dam foundation, the project
specifications required that the base of the cut-off trench
be located in "unrippable rock" i.e., rock that would
require special techniques such as blasting or hoe-ramming
for breaking and removal.

Typical subsurface conditions consisted of a residual
soil profile overlying relatively shallow phyllitic bedrock
with a transition zone from soil to rock. A typical
subsurface profile is shown in Figure 3. A wide range of
transitional materials existed between the friable soils
and indurated rocks. For most parts of the site, these
materials exhibited high strength and penetration

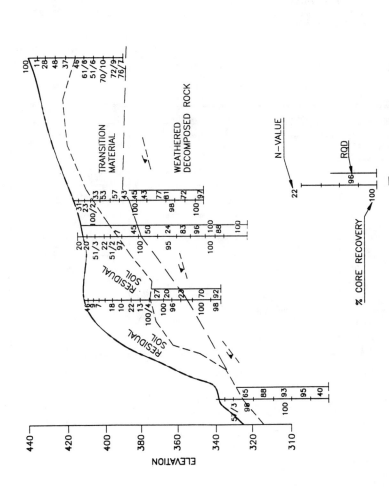

Figure 3. Typical Subsurface Profile: Excavation to Achieve Reduced Permeability

resistance and behaved more like rock than soil. During construction, these materials were difficult to excavate and typically required ripping to facilitate their removal.

The field investigation included drilling over one hundred borings, test pits, and conducting pressuremeter testing and geophysical surveys. SPT tests and rock coring, where rock was encountered, were conducted at most of the boring locations. Index and strength laboratory testing were also conducted on retrieved samples. Table 2 presents a summary of engineering properties for the in situ materials, as set forth in the geotechnical design report, based on the field and laboratory test results.

Table 2. Summary of Project Defined Material Properties

Property	"Soil"	"Transitional"	"Rock"
Penetration Resistance (bpf)	N<100, readily penetrated by hollow stem auger	N>100, likely to produce hollow stem auger refusal	N>100, requires rock bits or coring
Seismic Velocity (fps)	1000-5000	3500-8000	>6500
Weathering	Decomposed	Decomposed to Weathered	Weathered to Unweathered
Fracture Frequency (per foot)	N/A	2 to greater than 4	less than 3
RQD	N/A	0%-50%	30%-100%
Estimated Excavatability Characteristic	Routinely Excavated by Conventional Machinery	Difficult Excavation, Often Requiring Ripping or Blasting in Mass Excavation and Blasting in Trench Excavation	Typically Requires Blasting Prior to Excavation
Friction Angle (degree)	26-35	30-45	> 40
Cohesion (ksf)	0-10	5-50	> 40

The materials classified as "rock" were only those that could be sampled by NX core drilling. The portion of rock having a fracture frequency of 1 to 3 joints per foot and field permeability in the range of 10^{-2} to 10^{-4} foot per minute (fpm) was termed as "pervious" rock. The portion of rock containing few fractures, typically less than 1 joint per foot and field permeability on the order of 10^{-4} to 10^{-7} fpm, was termed "sound" rock.

In this case, the definition of rock was for the purpose of defining the relative impervious unweathered material. Estimation of the rock boundary was important because of seepage requirements. Design considerations called for the cut-off trench to be excavated to relatively impermeable material. With this criterion, indications are that the dam has performed successfully since the reservoir was filled approximately six years ago.

Situation 3: Excavation for Drilled Shafts

Drilled Shafts were proposed for the support of a ten-story office building in Columbia, Maryland. Typical diameters of the shafts ranged from 4 to 5 feet.

The site lies within the eastern Piedmont geologic region. The bedrock in the vicinity is classified as the Wissahickon formation. This formation consists primarily of metamorphic rocks, such as schists and gneiss. The surface soils were generally derived from the weathering of these metamorphic rocks and igneous intrusions.

Soil borings and SPT's were performed at 14 locations. Also, rock coring and pressuremeter testing were conducted at four locations. The borings were advanced until hollow stem auger refusal was reached. Rock coring was conducted thereafter.

The in situ residual soils graded with depth into a transitional material possessing characteristics of both soil and weathered rock. The thickness of this layer ranged from 14 to 20 feet with N-values of 10 to 30 bpf. Below the residual soils, the transitional materials graded with depth into decomposed and weathered rock. The boundary between the transitional materials and the decomposed/ weathered rock was as poorly defined as that between the residual soils and the transitional materials. N-values within the transitional materials ranged from 27 to 100 bpf. N-values for the decomposed to weathered rock, within the subsurface strata, exceeded 100 bpf.

For the upper 5 feet of the decomposed to weathered rock, core recovery ranged from 10% to 100% with RQD values of 0% to 80%. Pressuremeter modulus was estimated on the order of 2000 tsf for this zone. Below the top five feet of this material, weathered to unweathered schistose gneiss and gneiss was encountered. Core recovery values were on the order of 75% to 100% with RQD values of 30% to 80%. Figure 4 presents a typical profile of the subsurface conditions.

Design of the drilled shafts accounted for both end bearing and skin friction. An N-value of approximately 100/1" was chosen by the design engineer to represent the upper boundary of the material in which the drilled shafts were to be founded.

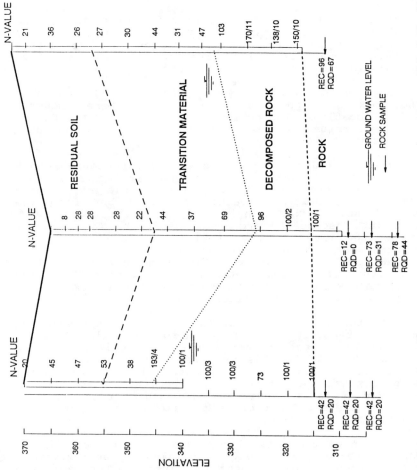

Figure 4. Typical Subsurface Profile: Excavation for Drilled Shafts

From an engineering perspective, the depth of embedment of the shafts was extended to where adequate capacity was to be maintained for the minimization of settlement. Therefore, the definition of the soil-rock boundary was perhaps not as crucial for the design as long as the design engineering properties were encountered in the field during construction.

From a construction perspective, the definition of the soil-rock boundary is directly related to excavatability. A truck mounted Hughes LDH rig was used for excavating the shafts. An earth auger was used to drill through the overburden soils. A single flight-helical auger bit with carbide tipped cutting teeth was used for advancing the hole through decomposed and unweathered rock. The criteria used by the contractor for estimating earth auger refusal was a penetration rate of less than 6 inches in 5 minutes.

Auger refusal, as defined by the contractor was reached near the boundary between the transitional material and the decomposed rock. This was well above the bearing material (having an N-value of 100/1") on which the design was based. Consequently, the contractor claimed additional "rock excavation" to achieve the design bearing elevations which, to the design engineer, represented top of rock.

In order to avoid the potential dispute, the subsurface materials in each shaft excavation were carefully logged to obtain site-specific information. An optimum length for each shaft, based on the actual conditions encountered in the field, was evaluated. Engineering design was adjusted accordingly.

Discussion

Several criteria exist to delineate the soil-rock boundary. Most of these criteria do not recognize the impacts of the specific engineering and construction applications on establishing such a definition. Every project is different and could have more than one functional definition for "rock", i.e., an engineering definition and a construction definition.

For the case studies presented above, common investigation methods used to characterize the subsurface material are summarized as:

Test Borings
o Standard Penetration Test
o Core Drilling

Seismic Refraction
o Seismic velocity

Direct Inspection
o Past projects in the area
o Test Pits and Trenches

Defining rock for projects where excavation must achieve an established grade is difficult. Excavatability is dependant on the material characteristics as well as type and conditions of the excavation equipment, the skill of the operator, and the available space for maneuverability. In relatively open excavations, a large dozer with a ripper can be used because it can maneuver. In this case, hard material can be excavated faster and in larger quantities. On the other hand, hoe-ramming or other rock removal technique would likely be needed in localized trenches and narrow excavations because of the limitation imposed by the working space on the excavation equipment.

From an engineering perspective distinction between, transitional material and rock is important in understanding the long- term behavior of the foundation material. Strength and physical properties, strike and dip of layers, joints and faults, and nature of the infilling material are engineering characteristics the knowledge of which is essential for competent design.

From a construction perspective, the definition of, and distinction between, transitional material and rock is crucial to avoid unpleasant surprises and lengthy litigation. The material substance (strong or weak); and the nature of the formation (massive, layered, blocky, or broken) are usually the main considerations for mass excavation and trenching.

Conclusions

This paper has investigated attempts to define the soil-rock boundary for different engineering applications in the Piedmont geologic provinces of the eastern United States. As discussed in this study, different criteria should be adopted according to the purpose and application for which the soil-rock interface characterization is required.

Geotechnical engineers, working with the owner and other members of the design team, usually establish excavation and foundation depths. These are functions of both the geometry of the facility and the demands for foundation support. In cases where building elements protrude into rock-like materials, the engineer utilizes the engineering characteristics of such materials to efficiently support the facility. The primary engineering concern is that the rock not be damaged or becomes more compressible during the removal of adjacent rock that protrudes into the profile of the facility.

Rock removal is a construction concern. About the most the design geotechnical engineer can realistically contribute to the evaluation of excavatability is to provide reliable test data such as blow counts and rock cores, compression wave velocities, etc. The contractor can use these data to estimate the excavatability characteristics of the in-place material. Few design engineers have the excavation experience to convert this data into reliable "rock" cost quantities for a construction bid. Such data correlation should be developed by the contractor who has knowledge of his own equipment, methods, and personnel. Also, the contractor can place a value on the time required to break and remove the rock with his resources.

Based on the experience of the writers, it is recommended that all excavation work should be bid on an "unclassified" basis with the following contract inclusions: First, the contract should provide the contractor with the available exploration data, test borings, laboratory tests, geophysical field data, etc. Furthermore, the contractor should be asked to base his bid on this baseline data plus any other data he may choose to acquire. The contract should contain a differing site condition clause indicating the contractor's right to rely on the provided data. Given these conditions, excavation contractors would have a clear mandate to do their job with the understanding that if the data are not representative of the site, they could then be paid accordingly.

References

Coates, D. F. (1970) "Rock Mechanics Principles," Department of Energy Mines and Resources, Mines Branch Monograph 874.

Deere, D. U., and Patton, F. D. (1971) "Slope Stability in Residual Soils," Proceedings of the 4th Pan-American Conference on Soil Mechanics and Foundation Engineering, Vol. I, pp.87-170.

Krukowski, J. (1990) "Can it be Ripped?, Seismic Analysis Aids Decision," Highway and Heavy Construction, December, pp 51-52.

Litke, S. (1991) "Rock Roundtable,"Article on ADSC January Meeting, ADSC Magazine, pp 31-35.

Obermeir, S. F. (1979) "Engineering Geology of Soils and Weathered Rocks of Fairfax County, Virginia," Open File Report 79-1221, U.S. Geological Survey, Reston, Virginia.

Peck, R. B. (1976) "Rock Foundations for Structures," Proceeding of the Specialty Conference on Rock Engineering, ASCE, Vol. II, Boulder, Colorado, pp 1-21.

Sowers, G. F. and Richardson, T. L. (1983) "Residual Soils of Piedmont and Blue Ridge," Transportation Research Record 919, pp 10-16.

White, R. M. , and Richardson, T. L. (1987) "Predicting the Difficulty and Cost of Excavation in the Piedmont," Proceedings, Foundations and Excavations in Decomposed Rock of the Piedmont Province, ASCE Geotechnical Publication No. 9, pp. 15-36, April.

Appendix I. Conversion to SI Units

To Convert	To	Multiply By
acre	hectare	0.4047
foot	meter	0.3048
kips	newton	4448.0
pound	newton	4.4480
cubic foot	cubic meter	0.0283
foot/sec	meter/sec	0.3048

GROUND STABILIZATION IN SHALLOW DEPTH
SOFT GROUND TUNNELING

Khosrow Bakhtar[1]
M. ASCE
William Martin[2]
Bruno Dietl[3]
Paul Byrnes[4]

Abstract

A new approach was employed for ground stabilization to drive a semi-circular cross-section tunnel with overall dimensions of 24 ft (7.3m) wide, 12 ft (3.7m) high, and 164 ft (50.0m) long. The tunnel was mined through sandy-silty-clayey soil, SM and SC based on the Unified Soil Classification System, with the invert located in thin-bedded claystone and soft to medium hard sandstone. The beds were typically highly folded and fractured, with occasional small faults. The tunnel under-crosses the Pacific Coast Highway, a major state freeway, with a maximum crown depth of 8 ft (2.4m) below the surface.

Location of the tunnel is within an active seismic zone, 1.2 miles (1.9km) from the Newport Inglewood Fault, which required seismic loads to be considered in design. The shallow depth of cover combined with the presence of a soil-rock interface along the tunnel invert presented a major challenge for the engineers involved to develop a design approach which would be feasible and cost-effective,

---

[1]President, Bakhtar Associates, 2429 West Coast Highway, Newport Beach, California 92663

[2]President, Mine Development and Engineering Corporation, 2421 Haley Street, Bakersfield, California 93305.

[3]Vice President, Valley Engineers, Inc., P.O. Box 12227, Fresno, California 93777.

[4]Director of Engineering, Coastal Community Builders - A Division of The Irvine Company, P.O. Box I, Newport Beach, California 92658.

while minimizing the risk of excessive settlement of the
road.   This paper describes the technique employed to
stabilize the surrounding soil mass and the method used to
design and construct the foundation footings at the soil-
rock interface.   Furthermore, the paper provides a
practical example on how to differentiate between soil and
rock for structural design purposes.

Introduction

An under-crossing for golf cart access was planned
under the Pacific Coast Highway, just south of the City of
Newport Beach in Orange County, California.   The tunnel
serves the Newport Coast Golf Course and allows golf carts
to cross the highway from the inland (north) side of the
highway to four holes at the ocean side.

Initial design of the tunnel was started based on
temporarily diverting each direction of the traffic and
constructing the tunnel using the cut-and-cover technique.
However, because of the excessive construction costs,
shoring difficulties, significant utility support problems,
and traffic delays which would have been entailed,
alternate methods of construction based on mining the
tunnel were sought.   Initial analyses based on elasto-
plastic and kinematic techniques, (Bakhtar, 1990),
indicated that large settlement followed by ground rupture
could occur if conventional soft ground excavation methods
such as shield tunneling were employed.   This was due to
shallow depth of cover and large dimensions of the tunnel.
The tunnel under-crosses a busy freeway, the cut-and-cover
approach was found to be expensive because of utilities and
shoring requirements.   Applications of chemical or
compaction grouting were investigated and not found to be
feasible.   Utilizing a shield would have required changing
the tunnel configuration to a circular cross-section, the
lower half of which would have to be backfilled.   This
would not have improved ground stability or minimized the
risk associated with the large settlement at the surface.

Following a careful evaluation of the geotechnical
information, a unique approach was developed to use
directional drilling to bore circular horizontal holes
above the tunnel crown footprint and subsequently pull 9 in
(23cm) diameter standard steel pipes, with perforations, to
provide a cellular arch above the tunnel for ground
stabilization prior to excavation.   The tunnel was then
excavated having a support arch above the crown protecting
the road.   The tunnel foundation consisted of two circular
steel casings, 3 ft (0.9m) in diameter, which were bored
and subsequently jacked in place using a conventional auger
boring-jacking method.   The presence of soil-soft rock
interfaces along the invert of the tunnel dictated the

choice for the diameter of the steel casing used for the foundation footings.

This paper outlines the method of analysis and approach used for ground stabilization which made the construction of the tunnel at the subject site possible.

Geotechnical Conditions

Field exploration and laboratory investigation along the alignment of the tunnel were performed by a local geotechnical firm (Leighton and Associates, 1990). The geotechnical investigation indicated that the site is underlain by bedrock of the Monterey Formation. The bedrock is overlain by a thick deposit of non-marine terrace materials. The bedrock consists of thin bedded orange-brown and gray clayey, very hard siliceous siltstone, and sandstone. The beds are typically highly folded and fractured, with occasional small faults. Several attempts were made to recover cores from the bedrock for mechanical testing. However, because of the highly fractured nature of the rocks, it was found impossible to acquire long enough samples suitable for laboratory characterization. A very poor rating was assigned to the bedrock based on the Rock Quality Designation (RQD).

Significant deposits of non-marine terrace materials occur, 10- to 20 ft (3.0 to 6.1m) thick, near the surface. The terrace deposits are crudely layered, with local layers having abundant amounts of angular siltstone, shale, and sandstone fragments, A 3- to 5 ft (1.0 to 1.5m) thick marine terrace deposit is preserved locally on the bedrock. These materials were deposited unconformably on an irregularly eroded wave-cut bedrock bench and are overlain by a non-marine terrace deposit. A typical profile from one of the exploratory holes is shown in the Appendix III. The groundwater level is well below the tunnel invert and did not effect the tunneling operation.

Design Load

The load on the tunnel support system is made up of static and active or dynamic forces. The dynamic component consists of traffic and earthquake loads. Figure 1 shows the tunnel dimensions used to calculate these loads.

The height of the load prism was calculated based on the approach outlined by Proctor and White (1977) using Equation (1)

$$H_r = 0.4 (b + H_t) \tag{1}$$

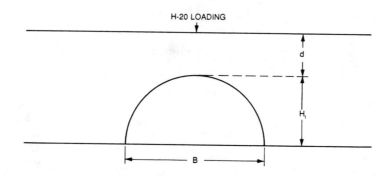

Figure 1.   Tunnel Dimensions [B = 24 ft (7.3m); H = 12 ft (3.6m); d = 8 ft (2.4m); $H_{20-Load}$ = 100 psf (4.8 kPa); ρ = 130 pcf (2.07 gm/cc)]

The overburden and live loads were calculated using Equations (2) and (3), respectively.

$$\sigma_v = \rho h_r \tag{2}$$

and

$$\sigma_{v1} = H_{20}Load = 100 \; psf(4.8 \; kPa) \; at \; 8\text{-}ft \; (2.4m) \tag{3}$$

and the total vertical stress

$$\sigma_{total} = \sigma_v + \sigma_{v1} \tag{4}$$

was calculated from Equation (4) to be 1,348 psf (64.5 kPa).

The safety factor for design of the steel support elements, wide flanges with liner plates, was calculated by incorporating the seismic load into the total load.   In calculating the seismic load, the horizontal component of acceleration was neglected -- the structure is flexible, located beneath the surface, and will tend to move with the overall ground motion.   The maximum vertical component of the acceleration was found to be 0.68g based on the local seismic records. The vertical component of the acceleration was added to the total load for design purposes. The resulting factors of safety were calculated to be 2.53 and 1.50 for total load and seismic load, respectively.

Determination of Footing Size

The load to be carried by each footing is the load exerted by each rib. Each footing carries the load equivalent to one-half of the tunnel width over a distance of 4 ft (1.2m) which is the rib spacing. Therefore,

$$P_{calculated} = P_{total} \cdot (B/2) \cdot d \tag{5}$$

In the above equation, the load (P) is expressed in lbs (kg); $B/2 = R =$ half tunnel width = 12 ft (3.7m); $d =$ rib spacing = 4 ft (1.2m); which results in the calculated load to be determined as 64,704 pounds (29,350 kg).

The mechanical and intrinsic material properties of soil and rock dictated the overall dimensions of the footings to carry the calculated load.

Because of the fractured nature of the bedrock, it was found impossible to recover cores long enough for material testing purposes. Therefore, soil properties were used to calculate the size of footings to support the tunnel load. Based on laboratory test results, the average shear strength of the earth materials encountered was characterized with an angle of internal friction of 29 degrees and a cohesion of 225 psf (10.8 kPa). The Terzaghi's equation for concentric loading on continuous footing was used to calculate the ultimate bearing capacity

$$q_{ultimate} = cN_c + \gamma DN_q + (\gamma B/2) N_\gamma \tag{6}$$

and yielded

$$q_{ultimate} = 6,300 + 1,920(D) + 780(B) \tag{7}$$

where D and B are depth and width of the footing, respectively.

Using the approach outlined in the Engineer Manual, (Waterways Experiment Station, 1969), the ultimate bearing capacity was calculated for a 3 ft (0.9m) diameter circular footing with a depth below lowest adjacent grade of 4 ft (1.2m) as:

$$q_{ultimate} = 6,300 + 1,920(4) + 780(3)$$

$$= 16,320 \ psf \ (781 \ kPa)$$

Similarly, for a 3 ft (0.9m) diameter 4 ft (1.2m) long footing, the load per unit area is:

$$\sigma_{3 \times 4} = \frac{64,704}{12} = 5,392 \ psf \ (258 \ kPa)$$

In the above calculation; the footing is circular, but with the long axis horizontal; therefore; the base of the footing is not planar.

The above approximation to determine the optimum diameter for the circular foundation is justifiable because once the support elements were erected, the entire assembly would act as a single unit resisting localized deformation.

As a result of local conditions and site specific soil characteristics, a decision was made to reduce the ultimate bearing capacity

$$q_{reduced} = q_{ultimate} \ (0.75) \tag{8}$$

$$= 12,240 \ psf \ (586 \ kPa)$$

The allowable bearing capacity was then calculated from

$$q_{allowable} = (0.5) \ q_{reduced} \tag{9}$$

The calculated load per square area for a 3 ft (0.9m) diameter footing was 5,392 psf (258 kPa) which is less than the allowable load of 6,120 psf (293 kPa). Therefore, a 3 ft (0.9m) diameter footing was found to be satisfactory.

Ground Stabilization

        The results of initial analysis (Bakhtar, 1990)
indicated that the excavation of tunnel using conventional
shield technique without stabilizing the soil above the
crown would lead to excessive settlement and possible
damage to the highway.   Compaction grouting was not
advisable because of the shallow depth of cover: it could
actually cause localized heaving or lifting of the highway.
Chemical grouting was not feasible because of the low
permeability of the soil.

        Following a thorough evaluation of the site
conditions, the design engineers decided on utilizing the
directional drilling technique to form a cellular arch
above the footprint prior to excavating the tunnel. Based
on this method, a series of horizontal holes were drilled
above the footprint of the tunnel and perforated steel
pipes (8 in (20.0cm) inside diameter, 0.5 in (2.3cm)
thick), referred to as the "spiles", were pulled through
these holes. Figure 2 shows the schematic of the tunnel
cross-section with the spiles and foundation casings in
place. The sequence of operation for spile installation
consisted of the following:

- a guided boring system, consisting of a
  trackdrill, was set at the elevation of each
  spile;

- machine was aligned and proceeded with the pilot
  bore;

- the first guided hole was bored using 1-1/2 in
  (3.8cm) drill rods;

- the drilled hole was enlarged prior to pulling
  the steel pipes using a 12 in (30.5cm) blade
  rimmer;

- the enlarged hole was cleaned using a packer
  rimmer;

- using the blade rimmer with a 10,000-lb (4,500
  kg) swivel, 9 in (22.8cm) OD perforated steel
  pipes were pulled through the rimmed hole.

        The steel pipes were delivered to the site in 20 ft
(6.1m) lengths and welded together prior to pulling them
through the holes. Following the completion of spile
installation, the 3 ft (0.9m) diameter steel foundation
casings were augered and jacked in place as shown in Figure
2.

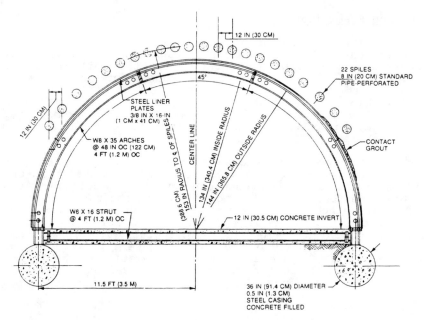

Figure 2.   Typical Tunnel Section

The spiles were filled with a special slurry-grout
at low pressure. The mix proportion of the slurry-grout
consisted of water, Portland cement (Type V), aquagel
bentonite, sand, and air entraining agent with the water-
to-cement ratio of 1.96. Low pressure applied to the grout
during pumping allowed penetration into the perforated
steel pipes (spiles) and formation of a support arch above
the tunnel footprint. The installation of the spiles
enabled settlement of the highway to be controlled within
the 1 in (2.5cm) limit.

The large diameter steel casings at the foundation
locations were filled with conventional concrete. While
the work was in progress for spiles and foundation casings,
the excavation for the portal and construction of the
headwalls at the ocean side were completed.

Remarks

The preceding sections of this paper describe the
stabilization technique and foundation design methodology
employed for construction of the subject tunnel. The

approach was unique in terms of utilizing directional drilling for spile installation to protect against excessive settlement and in the design of foundations. The new ideas associated with installations of spiles and circular foundation footings enabled tunnel construction to proceed without a shield and settlement to be controlled within 1 in (2.5cm).

The successful design and construction of the subject tunnel provides answers to some of the questions of interest to engineers for differentiating between soil and rock. Locally, in Southern California, most geotechnical engineering firms are staffed with engineers having expertise in soil mechanics. Information on soil properties is commonly available or can be obtained from bucket-auger borings at any specific location. However, most of the local experience is in the area of foundation design for buildings where a factor of safety of about three is required to minimize post-construction problems arising from excessive settlement of the structures. Furthermore, around the coastal areas very low values are assumed for soil cohesion and angle of internal friction. For the tunnel under consideration, the invert and foundation footings were located in soft to medium-hard siltstone and sandstone overlain by marine and terrace deposits. The soft underlain rock materials certainly possess cohesion and internal friction which should be accounted for to design the foundation footings. This circular footing foundation was bored and subsequently pulled in place. A safety factor of three is totally inappropriate because once the steel ribs and invert struts are installed, the whole structure can not undergo any displacement relative to the surrounding ground (refer to Figure 2). For the subject tunnel, allowing reasonable values (based on laboratory characterization) for cohesion and the angle of internal friction of the soft underlying materials (soft rock) and successful implementation of spiles enabled the proper economic design and safe construction of the structure to be achieved.

Acknowledgement

Funding for the design and construction of the tunnel was provided by The Irvine Company. Valley Engineers, Inc., was the prime contractor for the construction. Design plans and specifications were prepared by Mine Development and Engineering Corporation in association with Bakhtar Associates and Valley Engineers. T. G. McCusker was a design consultant. Geotechnical investigations were performed by Leighton and Associates. Bakhtar Associates provided technical consulting for the project during the design and construction phases of the contract.

Appendix I. References

Bakhtar, K., "Analysis of Excavation Induced Ground
Deformation at the Proposed County of Orange Golf Cart
Tunnel", Bakhtar Associates, Internal Report 90-05,
Submitted to The Irvine Company, July 1990.

Leighton and Associates, "Foundation Investigation for the
Proposed Golf Cart Undercrossing", Project No. 18413172,
February 1990.

Proctor, R. V., and White, T. L., "Earth Tunneling with
Steel Support", Commercial Shearing Stamping Company,
Pennsylvania, 1977.

Waterways Experiment Station, "Engineering and Design -
Conduits, Culverts, and Pipes", Engineer Manual, EM 110-2-
2902, Department of the Army, March 1969.

Appendix II. Notation

The following symbols are used in this paper:

B	=	tunnel width
c	=	cohesion
d	=	overburden depth
H_t	=	tunnel height
H-20 Loading	=	traffic load
P	=	load
ρ	=	density
H_r	=	height of load prism
σ_v	=	overburden stress
σ_{v1}	=	live stress
P	=	load
N_c	=	28, bearing capacity factor
N_q	=	16, bearing capacity factor
N_γ	=	13, bearing capacity factor
γ	=	unit weight
q	=	bearing capacity
OC	=	off center
OD	=	outside diameter

Appendix III. Geotechnical Boring Log

GEOTECHNICAL BORING LOG

SHEET 1 OF 2

PROJECT NAME	Cameo - PCH Tunnel			
PROJECT NO	1841372-14		BORING DESIG.	SB-12
DATE STARTED	8/18/90	DATE FINISHED 8/18/90	STATION	
DRILLER	Pioneer Drilling	LOGGED BY MMM/TST	OFFSET (FT)	
GROUND WATER ELEV	91.0 +	GW DEPTH (FT) 41.0	GSE (FT)	132.0 +
TYPE OF DRILL RIG	Mobile B-61	DRIVE WT (LBS) 140 lbs.	DROP (IN)	30"

DEPTH (FEET)	ELEV	SAMPLE TYPE	SAMPLE	BLOWS/FT OR REC./RQD	GRAPHIC LOG	ATTITUDES	GEOTECHNICAL DESCRIPTION	MOISTURE CONTENT%	DRY (PCF) DENSITY	SHEAR STRENGTH (KSF)	OTHER TESTS
							Surface: Brown clayey fine to coarse SAND, damp (FILL).				
	130	D	1	57			**TERRACE DEPOSIT (Qtm)** @2': Dark brown clayey fine to coarse SAND with roots and some gravel, dense, damp to moist (SC).	16	112		DS
		B	1								MD,AL GS
5		D	2	55			@5': Mottled orange-brown and gray clayey fine to coarse SAND, dense, damp to moist (SC).	10	114	2.0	UU,PM
		S	1	30				10			SC
	125										
		D	3	44			@8': Orange-brown clayey fine to coarse SAND, dense, moist (SC).	9	112	1.6	UU,GS
10		S	2	33			@10': Same.	9			SC
	120	D	4	42			@12': Orange-brown clayey fine to coarse SAND, dense, moist (SC).	9	106		DS
		D	5	58						1.8	
15								21	100		
		B	2				**MONTEREY FORMATION (Tm)** @15': Bedrock: Mottled orange-gray silty SANDSTONE, dense, damp, (SM).				MD,AL SC SC
	115	S	3	34			@16.5': Mottled orange-gray and brown fine sandy CLAYSTONE with some silt, dense, moist (CL).				
		D	6	70			@18': Mottled orange-gray and light brown fine sandy CLAYSTONE with 10mm fine sand lenses, dense, moist (CL).	24	99		DS UU
20		D	7	43			@20': Same.			2.4	UU,AL GS
	110	S	4	69				10	114		
							@22.5': Hard Drilling. @23': Mottled light brown and black clayey fine SANDSTONE, very dense, damp (SC).	13			
		B	3								SC
25											
	105						@27': Light brown CLAYSTONE with trace of fine SANDSTONE and 2mm lenses of black fine sand, very stiff, damp (CL).	12	113		
		D	8	92							
30											

SAMPLE TYPES:

C ROCK CORE	B BULK SAMPLE	GW WHILE DRILLING		
S SPLIT SPOON	T TUBE SAMPLE	GW HRS.	C CONTACT	
D DRIVE SAMPLE	b SMALL BAG	BEDDING PLANE	F FAULT	
		JOINTING	S SHEAR	

LEIGHTON AND ASSOCIATES INC.

"Rock" in the Limestone Regions of the Southeastern
United States

E. C. (Buddy) Yokley, Jr.[1]

Abstract

Problems often arise in the shaft drilling industry in
the Southeast. The karst areas in this region have certain
situations found in few other places. Erratic boulders and
limestone pinnacles often test the definition of "what is
rock?" for pay purposes.   A thorough subsurface
investigation can prevent many unfortunate situations from
developing and assure timely completion of foundation
drilling projects.

Introduction

Through the years, discussion on "where does rock
begin," have produced many varied and sometimes
controversial opinions. These opinions vary from region to
region and from formation to formation. For purposes of
discussion, this paper focuses on those limestone
formations found in the southeastern United States west of
the Appalachian Mountains, Fig. 1. The author's extensive
experience in the foundation drilling industry has led to
some observations not readily available to the design
engineer.

One of the most widely accepted definitions of rock is
published in the Standards and Specifications for the
Foundation Industry (Association of Drilled Shaft
Contractors, 1991):

> Rock is defined as any material which cannot be
> drilled with a conventional earth auger and/or
> underreaming tool, and requires the use of
> special rock augers, core barrels, air tools,

---

[1]Vice-President, Long Foundation Drilling Company, P.O. Box
266, Hermitage, TN 37076-0266

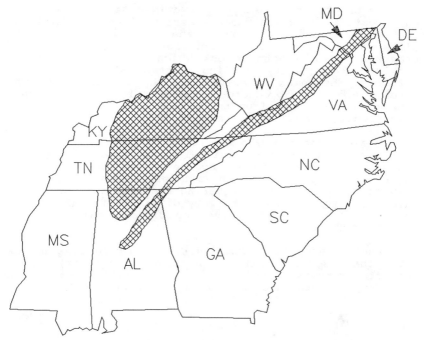

Figure 1.   The Karst Regions of the Southeastern United
States

blasting   and/or   other   methods   of   hand
excavation.   All earth seams, rock fragments,
and voids included in the rock excavation areas
will be considered rock for the full volume of
the shaft from the initial contact with rock for
pay purposes.

This definition is very reasonable and practical
for application to work in most limestone areas.   However,
when the degree of weathering in some formations produces
a very erratic top-of-rock profile, determining "where does
rock begin?" for pay purposes becomes difficult.

Problems with the Definition of Rock

One difficulty with the definition as given by the
Association of Drilled Shaft Contractors (1991) is found in
the last sentence.   Once rock is encountered, payment for
rock excavation will continue to the bottom of the shaft.

This is usually the case, but sometimes isolated boulders can create a situation where a serious inequity develops. The purpose of the wording in the rock definition is to allow continuous rock payment for layered, seamy, or otherwise unsatisfactory rock which must be removed to provide the quality bearing surface desired by the structural engineer. However, if the weathering of the parent bedrock has produced boulders, "floaters," or other rock masses detached from the parent formation, obviously this situation must be addressed.

One example of this condition is as follows. After drilling through the initial soil overburden and encountering and removing a boulder, perhaps 0.6 m (2 ft) thick, drilling continues in soil for another 2.5 m to 3.0 m (8 ft to 10 ft) before entering rock again, Fig. 2. In this situation, it would be very difficult to make a case for payment of rock from the top of the boulder to the bottom of the shaft.

Another example occurs when drilling refuses on a rock layer that is approximately the same 0.6 m (2 ft) thick, but after removal of this layer, a six inch seam of earth is found, beneath which rock is again encountered, Fig. 3.

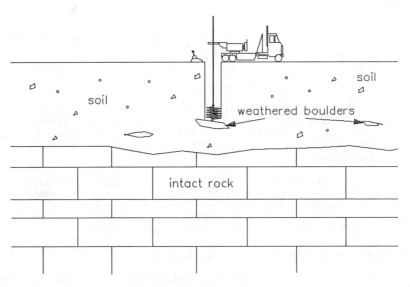

Figure 2.   Limestone "Floaters" Within The Soil
Overburden

It is not unusual for this condition to continue for depths of 1.5 m to 6.1 m (5 ft to 20 ft) or more, depending on the formation involved. In this case, the drilling contractor could not be expected to issue credit for these dirt seams since he would be constantly changing tools, taking measurements, etc. A state of chaos would result.

So, where can a line be drawn? It is difficult to determine at best and many attempts at such a determination have been made. Long Foundation Drilling has, on occasion, agreed to accept a vertical dimension of 0.9 m (3 ft) for use under these circumstances. An agreement in the field should be reached that would provide for credit to be issued at an earth credit unit price if soil seams or layers were drilled which exceeded 0.9 vertical linear meters (3 ft). For seams of 0.9 m (3 ft) or less there would be no credit allowed and the entire depth would be paid as rock excavation. An agreement such as this would make provisions for an equitable adjustment, when rock is underlain by a substantial soil layer, prior to reaching competent bearing material. It is not intended, however, as a means by which to "nit-pik" the drill footage, and could not be considered applicable in either shot rock fills or highly laminated rock formations. In shot rock

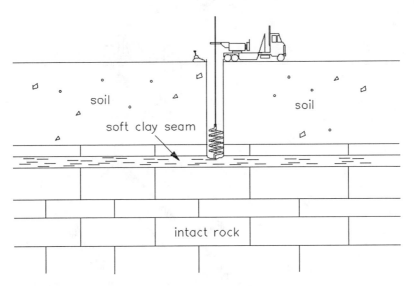

Figure 3. Thin Seams of Soil Can Cause a Change in Conditions

fills, the rock boulders and fragments are placed in
continuous layers. Since the rock is not in a solid layer,
rock tools must be used entirely for removal. In highly
laminated formations, the rock and soil layers alternate so
frequently that a determination cannot be made until a
significant rock layer can be reached.

Another "where does rock begin?" situation often
occurs in pinnacled limestone formations. When hard
limestone rock dips at a very steep angle from the
horizontal, the earth auger will tend to deflect due to its
inability to penetrate the rock mass, Fig. 4. If drilling
is allowed to proceed, the shaft will lead off from its
vertical alignment and an "out-of-plumb" shaft will result.
For this reason, the shaft excavation in earth must be
halted immediately to preserve the integrity of the shaft.
At the point where deflection begins, the top of rock is
established. This condition will also become apparent,
when during placement of the temporary casing, it becomes
impossible to advance the casing further without the use of
twisting or coring procedures.

In summary, whenever conventional soil drilling
methods can no longer advance the shaft excavation without

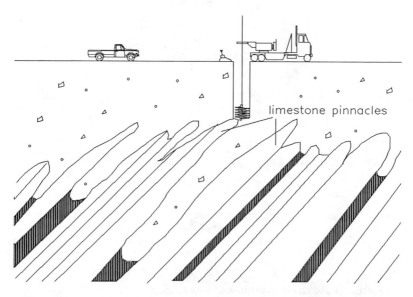

limestone pinnacles

Figure 4. Limestone Pinnacles Can Deflect the Auger

meeting either refusal or deflection from the intended
alignment, rock excavation methods must be initiated and
payment for rock excavation will become effective at that
elevation.

Understanding the Definition of Rock

Lack of proper understanding on "where does rock
begin?" often creates unanticipated overruns for the owner
and hostile working conditions at the job site. On a
recent project in eastern Tennessee, drilled shaft rock
excavation quantities exceeded the proposed budget to such
an extent that the building had to be completely
redesigned. After completion of only a few drilled shafts,
it became apparent that the amount of rock excavation which
had been included in the base bid was only a fraction of
the amount required to achieve satisfactory bearing.
Needless to say, the project was halted and the owner was
extremely unhappy and dissatisfied. Several floors had to
be omitted in order to sufficiently reduce the loads to the
point where they could be supported by the use of spread
footings. A situation like this can, and should, be
avoided by taking proper precautions in the preliminary
stages of development. Determination of "where does rock
begin?" once again plays an important role.

The most effective way to prevent unexpected cost
overruns due to unanticipated rock excavation is for the
engineer to carefully evaluate the subsoil information, and
have a thorough understanding of how rock is measured.
Careful preliminary planning by the engineer can reduce the
risk of controversy on drilled shaft projects requiring
limestone excavation by incorporating the following steps:

1. Adequate subsoil information. Be sure the owner
    authorizes sufficient soil and rock exploration.
    Coring of rock is essential to determine the
    extent of weathering.

2. Understanding the definition of rock. Be sure
    the engineer and the drilled shaft contractor
    have the same interpretation of the
    specifications.

3. Classification of drilling quantities in the bid
    documents. Unclassified excavation in drilled
    shaft work will often result in heavy bid
    "padding" by the drilled shaft contractor. He
    may assume a defensive position and an attitude
    that the owner will "pay me now or pay me
    later." When the owner is willing to share in

this risk, he will receive a more realistic bid.

4.   Inclusion of rock quantities which represent the
      very best estimate of anticipated rock removal.
      When this is done, "game playing" by both the
      engineer and the contractor is reduced or
      eliminated, and the owner once again becomes the
      beneficiary.

When these steps are utilized, fewer surprises and/or
controversies will be encountered during construction.

Conclusion

      Education and cooperation will always provide the best
results when dealing with difficult and unknown conditions.
A "good deal" is really only a good deal when it is good
for both parties. The exchange of information between the
engineer and the drilled shaft contractor can provide
useful information during the development of a drilled
shaft foundation design. The dreaded "where does rock
begin?" definition can become a relatively simple and
routine determination, by understanding and utilizing
standardized drilled shaft terminology and practices. It
is very important that they are fully understood by all
concerned parties. The most successful projects are
usually the result of the greatest cooperation among the
participants.

Appendix. Reference

Association of Drilled Shaft Contractors (1991). Standards
and Specifications for the Foundation Industry, Association
of Drilled Shaft Contractors (ADSC), P.O. Box 280379,
Dallas, TX, 75228.

The U.S. Office of Surface Mining (OSM) Proposed Strength-Durability Classification System

Robert A. Welsh, Jr.[1]
Luis E. Vallejo, Member ASCE[2]
C. W. Lovell, Life Fellow ASCE[3]
Michael K. Robinson[4]

Abstract

Federal regulations for the surface mining of coal stress proper design and construction of stable waste rock disposal areas. A key to the long-term stability of these excess spoil fills is accurate characterization of rock strength and durability. The present study of 115 bulk overburden samples collected from 61 mine sites was designed to develop testing methods and a classification system that will more accurately identify weak, non-durable Appalachian rock in a timely and cost-effective manner. The Office of Surface Mining (OSM) rock strength-durability classification identifies rock that is likely to degrade to finer material in durable rock fills and rock drains when subjected to gravity loading by the fill mass and swelling strains induced by infiltrated water.

Introduction

Durable rock is defined as rock which does not slake in water and will not degrade to soil material by 30 CFR §816.73(b) (United States Code of Federal Regulations, 1989). The intent of this durability standard is to selectively obtain rock that can withstand surface mining

----

[1]Geologist, U.S. Office of Surface Mining, 1020 15th St., Denver, CO 80202.
[2]Civil Engineer, U.S. Office of Surface Mining, Ten Parkway Center, Pittsburgh, PA 15220, and Associate Professor of Civil Engineering, University of Pittsburgh 15261.
[3]Professor of Civil Engineering, School of Civil Engineering, Purdue University, West Lafayette, Indiana 47907.
[4]Supervisory Physical Scientist, U.S. Office of Surface Mining, Pittsburgh, PA 15220.

conditions without significant degradation, particularly those affecting the fill mass after final placement. The objective of the present study was to select rapid, inexpensive strength-durability testing standards which clearly differentiate between strong-durable and weak or nondurable materials. Accurate classification of spoil will provide better data with which to model the long-term stability of excess spoil fill structures.

The rock samples tested in this study were collected at recently blasted highwalls of 61 surface mines in steep slope areas of Kentucky, Tennessee, Virginia, and West Virginia. A total of 115 grab samples of freshly blasted rock, each weighing approximately 445 N each, were collected at each sample site. Early to Middle Pennsylvanian-age rocks associated with presently-mined coal beds were sampled for this study. The 115 rock samples were classified as follows: 78 shales, 8 mudstones, and 29 silt- or sandstones.

Rock Strength-Durability Analyses

The final test program that evolved for use in this study included the following tests:

a) Jar Slake (Soak) Testing provides a qualitative measure of rock durability behavior after immersion in water for a 24-hour period. This test is particularly relevant to the durability of rock drains in valley fills, as suggested by Andrews et al. (1980) who state, "...this test might be most closely related to spoil materials located at depth and within a constant humidity or totally saturated environment." The sample is immersed in a glass container holding tap or distilled water at 20°C, and observations of any disaggregation are made. The relative degree of slaking can be ranked using a description and photographic scale developed by Lutton (1977).

b) Free Swell Index Testing indicates the slaking stress that affects rock when it is removed from its in-situ environment in the overburden, and is exposed to moisture. This is done by measuring the volume expansion of the cored rock normal to bedding upon immersion in water for a period of 12 hours immediately after oven drying to 105°C. A dial gage is used to measure any dilatancy. Free swelling strains are probably the result of expansion due to air pressure-induced breakage along interconnected voids such as microcracks in the rock (Olivier 1979).

c)   Point Load Testing also simulates loading stresses
     by inducing tensile failure in rock samples. The
     point load device employs two conical platens to
     apply the load. Twenty rock lumps are loaded to
     failure, and the median corrected point load
     strength index is calculated following the procedure
     outlined by Oakland and Lovell (1982).

The above tests were conducted by the U.S. Army Corps
of Engineers Missouri River and South Atlantic Division
Geotechnical Engineering laboratories under the testing
protocol designed by OSM.

The OSM Rock Strength-Durability Classification System

The OSM system for classification of rock strength
and durability consists of a two-phase testing protocol
(fig. 1). The first phase is durability testing by the
jar slake test. Those rock samples exhibiting significant
slaking by disaggregation or breakage into fine particles
are considered nondurable and are not suitable for durable
rock applications on the minesite. Rocks that pass this
first testing phase are subjected to a second testing
phase to certify the rock as of sufficient strength and
durability for rock drain or end-dump fill applications.

This second testing phase employs a two-index
strength testing procedure consisting of free-swelling
index testing of samples immersed in water for a maximum
of 24 hours, and point loading of rock samples.

Phase One Testing: Jar Slake Test

The OSM (United States Code of Federal Regulations,
1989) recognized the role of infiltrated water in rock
slaking in the permanent program regulations concerning
excess spoil fills at 30 CFR §816.71(f)(3), and in the
durable rock fill regulations at 30 CFR §816.73(b).
Durable rock is specified as rock that "...does not slake
in water..." Preamble comments pertaining to these
regulations indicate that this performance standard was
adopted to "...insure that subdrain material be
sufficiently durable to prevent degradation which could
result in blockage of the drain and subsequent failure of
the fill." (United States Federal Register, 1979).
Further, unanticipated settlement caused by consolidation
of large volumes of nondurable rock can disrupt the
surface drainage system of fill structures (Kirk and Hall,
1982).

Jar slake or soak testing is described by Andrews et
al. (1980) as "...a useful and simple tool for assessing

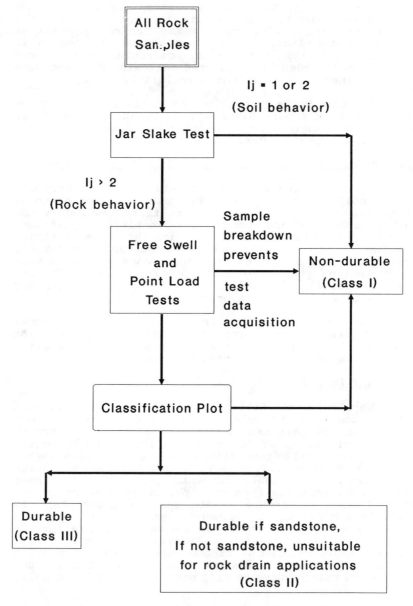

Figure 1. Strength-Durability Classification Flowchart

the durability of geologic materials." Strohm et al. (1978) recommend the jar slake test as the "...basic screening test..." for nondurable shales. The jar slake test provides such quick, inexpensive results that Lutton (1977) suggests that it could be used for continuous characterization of material. Venter (1980) concluded that "visual examination of wetted and dried samples (as in the jar slake test) may be more useful" than slake durability index testing (Franklin, 1970) of disintegrating mudstones.

Jar slake data are qualitative, and based on an assessment of the behavior of an as-received or oven-dried sample after a specified period of immersion. A ranking system based on the appearance of the sample after soaking is described by Lutton (1977). The jar slake index (I_J) is ranked on a scale from one to six, as follows:

I_J	Behavior
1	Degrades to a pile of flakes or mud
2	Breaks rapidly and/or forms many chips
3	Breaks slowly and/or forms few chips
4	Breaks rapidly and/or develops several fractures
5	Breaks slowly and/or develops few fractures
6	No change

The Lutton (1977) ranking system was applied to rock samples in this study. Irregularly-shaped samples weighing at least 100 grams at as-received moisture contents were soaked in distilled water at 20°C for 24 hours, and then photographed. Pre- and post-test photographs were compared to assess the mode and degree of breakdown.

The full range of slaking behavior, from I_J=1 to I_J=6, was observed in the tested rock. Of the 106 samples tested by the jar slake method, 24 samples were rated at an I_J of either 1 or 2. These samples were all fine-grained shales or mudstones. The mode of slaking observed was primarily chipping or flaking of the rock to form a loose pile of fine material. Strohm et al. (1978) recommend that "shales with a jar slake index, I_J, of one or two obviously should be considered soillike (sic) without further testing." These rocks would not meet the regulatory criteria for rock drain material mandated by 30 CFR §816.71(f)(3) or durable rock by 30 CFR §816.73(b), which require that rock in these applications do not slake in water and will not degrade to soil material.

Lutton (1977) indicated that damp material was generally insensitive to the jar slake test, compared to oven-dried rock. The rock sampled for this study was tested at the as-received moisture content, and thus slaking behavior may have been attenuated. The OSM-recommended procedure for jar slake testing includes initial oven-drying.

Phase Two Testing: Free-Swelling Index Testing

The effect of water on the swelling behavior of shales and mudstones has been documented by various authors, and summarized by Olivier (1979). Long-term strain measurements indicate that anisotropical, differential swelling and shrinkage strains develop in low-durability shales and mudstones when subjected to the following conditions:

a)   A continuous variation in the environmental humidity (Van Eeckhout, 1976)

b)   Alternating wetting and drying cycles of variable intensities and duration (Grice, 1968; Ollier, 1969)

c)   Prolonged periods of intensive air drying, followed by moisture absorption (Olivier, 1979)

These conditions are likely to be encountered by rock situated in excess spoil fill rock drain locations. Swelling of shale as a result of water absorption will cause a decrease in the shear strength of the rock (Bjerrum, 1967). This loss of shear strength is not only significant in rock drains, but is particularly important in the critical toe area of the fill, which is instrumental in maintaining the stability of the structure. Further compounding these concerns is the observation that the area most conducive to swelling strain generation is the toe of the fill, because this area has less gravity loading due to decreased cover and consequently, relatively low confining pressures (Bjerrum, 1967; Quigley et al., 1971; Holtz, 1959). Hoek and Bray (1977) credit Terzaghi with the observation that the lowest factor of safety for circular analysis is produced by a failure surface projected through the toe of a fill mass, assuming that the angle of internal friction (ϕ) of the material is greater than five degrees. As typical mine spoils have ϕ values in excess of 30°, the toe region is the most critical in terms of excess spoil fill stability.

Swelling and the resulting loss of rock integrity often occur without immediately visible signs of slaking such as cracking, sloughing, or dispersion (Moriwaki, 1974; Seedsman, 1986). Olivier (1979) observes that swelling strains are often anisotropically distributed, leading to localized areas of rock mass failure. As Seedsman (1986) indicates, such localized failures may not be immediately apparent as slaking.

Free-swelling behavior during water immersion was measured for 111 samples collected in this study. Swelling behavior ranged from zero volume expansion in some non-swelling sandstones to disintegration of mudstones that was so complete that an accurate dilation could not be measured. As Duncan (1969) observes, this disaggregation indicates that the tensile stresses generated within the rock exceeded its resistance to such forces. A total of 36 shales or mudstones had a swelling coefficient (ϵ_D) of greater than 0.02, or two percent. Only thirteen of these samples had an observed I_J of 1 or 2 (behave as soil), suggesting that the swell test is more sensitive to this type of water-induced slaking than the simpler jar slake test.

Olivier (1979) reports that at least 75% of the maximum free-swelling occurs within two to four hours after the immersion of the oven-dried rock sample in water. Nascimento et al. (1968) found that free-swelling of metamorphosed rock samples had substantially concluded after two to five hours. These conclusions are supported by swelling measurements for the present study. Free-swelling rates measured for ten shale and mudstone samples indicate that, by the end of four hours, an average of 90.8% percent of the ultimate swell magnitude had occurred. The percentages ranged from 80% to 99.1% of the ultimate swell.

For Appalachian shales or mudstones, a factor of 1.1 can be multiplied by the four-hour free-swell value to normalize these results to closely approximate the 12-hour free-swelling index. This revision to a shorter saturation period was suggested by Olivier (1979) with the caveat that such an abbreviated testing period is conditional on the time-dependent swelling behavior of the rock tested. This shorter testing period of four hours appears warranted for the rocks tested in this study, as typical free-swelling values plot as a nearly horizontal line after approximately two hours.

The OSM-recommended free-swelling test method follows procedures suggested by the International Society for Rock Mechanics Committee on Laboratory Tests (ISRM, 1979).

This method is based on research reported by Duncan et al.
(1968), Duncan (1969) and Nascimento et al. (1968). The
OSM method includes some minor modifications incorporating
observations of swelling rates and orientations made
during the present study. Free-swelling dilatancy is
measured along the axis perpendicular to bedding planes
(Z-axis), as the free-swelling rate and magnitude has been
observed to be several times greater in this direction
than parallel to bedding (Olivier, 1979 and Lo et al.,
1978). This result was confirmed during this study where
Z-axis swelling was typically 4 to 5 times greater than
the swelling observed along the X- and Y-axes. This
behavior results from preferential separation along
bedding plane laminations. For shale lithologies with
pronounced anisotrophy due to closely-spaced bedding
laminations, this measurement provides data on the maximum
swelling potential independent of any ultimate rock
orientation in the rock fill or backfill. This is also
the most common mode of breakdown in shales.

The rock specimen used for free-swell testing should
be derived from intact core, or sawed from a bulk sample
into a rectangular prism. The Z-axis must be oriented
perpendicular to the bedding laminations in order to allow
measurement of the maximum free swell. The minimum
specimen dimensions should be at least 15 mm or ten times
the maximum grain diameter, whichever is greater (Lama and
Vutukuri, 1978). An NX-diameter rock core is generally
satisfactory for direct testing purposes. The suggested
standard procedure for conducting the test follows that
described in ISRM (1979), incorporating an abbreviated
testing period of two hours, based on results from this
study.

Investigation of clay mineral content for the
Appalachian shale and mudstone samples tested in this
study indicate the operation of a swelling mechanism that
is not dependent on the presence of swelling clays. The
X-ray diffraction analysis of clay species in these
lithologies indicated that relatively insignificant
amounts of swelling clays were present. The most
prevalent clay minerals were the relatively non-swelling
species kaolinite and illite. The non-swelling clay
behavior is further substantiated by the generally low
plasticity indices (ranging from 4 to 27.7) from Atterberg
limits testing of fine rock particles. Swelling behavior
generally increases with increasing plasticity values
(Hart, 1974).

Considering the above evidence, the likely mechanism
of slaking for the shales and mudstones rocks is the
physical process of air pressure-induced breakage along

microcracks and other voids described in previous research by Terzaghi and Peck (1967) and Olivier (1979). Moriwaki (1974) concluded that nonswelling clay minerals are comparatively inert and do not accommodate internally-generated stresses that lead to pore air breakage. Pore air breakage occurs through development of high internal capillary suction pressures during the immersion of rock in water or exposure to a humid environment after periods of drying (Olivier, 1979). As stated previously, these conditions would be similar to those encountered in excess spoil fill rock drains.

Phase Two Testing: Point Load Testing

Rock strength measurements have been used by many investigators to classify rock for geotechnical engineering applications. Various strength tests, including uniaxial and triaxial compressive strength, point load, Schmidt rebound hammer, and hardness tests, have been used in rock classification systems for engineering applications (Deere and Miller, 1966; Franklin, 1970; Augenbaugh and Bruzewski, 1976; Strohm et al., 1978; Olivier, 1979).

Triaxial compressive strength tests are time-consuming, sophisticated, and expensive, and therefore do not meet several criteria for the present research. The Schimdt rebound hammer is difficult to use with soft materials such as shales and mudstones because of their tendency to fail under impact (Aufmuth, 1974). Because these lithologies are commonly encountered in the surface mining of coal, this device would not be appropriate. Hardness tests are similarly difficult to use with these soft rocks and were also rejected. Uniaxial unconfined compressive strength tests are laboratory tests only, and involve significant sample preparation prior to testing.

Point load testing data can be readily collected in the field, and have been successfully correlated with compressive strength results. Sample preparation requirements are minimal.

Broch and Franklin (1972) and Bieniawski (1975) have proposed use of the point-load strength test as an indirect way to obtain the uniaxial compressive strength of rocks. According to Broch and Franklin (1972), the advantages of using the point-load strength test are:

a) specimens in the form of irregular lumps are used and require no machining

b) smaller forces are needed so that a small,

portable testing machine can be used

c)   fragile and broken materials can be tested

d)   the test is inexpensive

The point load apparatus compresses a piece of rock between two points using two cone-shaped platens.   The rock piece of maximum dimension, D, is compressed diametrically to failure under a point load P.   From the test, a point load index, I_s, is determined as:

$$I_s = P/D^2$$

To standardize the index, I_s, it is converted to an equivalent value for a 50 mm diameter rock core (NX core = 54 mm in diameter) (Broch and Franklin, 1972).

The point load index, I_s, for a sample of shale or sandstone was calculated as the mean of 20 values of I_s determined from testing 20 lumps of the same rock sample in the point load apparatus. According to Oakland and Lovell (1982), 20 lumps is the minimum number that should be   tested   in   order   to   obtain   a   statistically representative value of I_s for one rock sample.   This reflects the ISRM (1979) recommendation that "...more (than 10 tests should be run) if the sample is heterogeneous or anisotropic" (statement in parentheses added). As already described, the rock types encountered in   coal-bearing   rocks   are   often   heterogeneous   and anisotropic by reason of depositional and compositional factors.

The rock samples were tested in the point load apparatus under a constant rate of strain of 0.025 cm/min. In addition, each of the determined I_s values was normalized to correspond to a 50 mm rock core using the charts published by Broch and Franklin (1972).

The shales and the sandstones were tested in the point load device at as-received and "saturated" moisture contents. Samples were considered "saturated" after water immersion for a period of 24 hours.

OSM Strength-Durability Classification (Swell Test-Point Load Test Plot)

Olivier (1979) devised a dual index "Geodurability" classification system for fine-grained sedimentary rock that employed free-swell tests in concert with uniaxial compressive   strength   testing.   He   suggested   the possibility of substituting the point load test for compressive strength testing to provide a measure of rock strength. The recognized linear relationship between these two tests for Appalachian coal measure rocks, as confirmed in Vallejo et al. (1989), serves to validate this concept. The point load and free swelling indices are therefore adopted as the standard indices of the final testing phase for OSM rock drain strength-durability determination.

As shown in figure 2, a plot is constructed with normal logarithmic scales comprising the X- and Y-axes. The range of free-swelling coefficient values, ϵ_D, from 0.0001 to 0.10 is plotted along the Y-axis. Point load values from 0.1 MPa to 10 MPa are plotted along the X axis.

Comparison of plots of free swelling and point load data with jar slake test data have suggested boundaries that may be utilized to create three engineering use categories. These classes are plotted on the chart (Figure 2) and defined as follows:

Class I-<u>Nondurable and weak rock</u> is not suitable in any <u>durable</u> rock application in excess spoil fills. This rock may comprise the 20% by unit volume nondurable component of the federally-defined durable rock fill mass. This material:

    a) behaves as soil in the jar slake test

    b) fails under sample preparation for either the free-swell test or point load test

    c) has a measured strength of less than 2 MPa, or a free swell of greater than 4%

Class II-<u>Rock that is suitable for rock drains only if sandstone, if rock is not sandstone, then it is</u> <u>**not** suitable for rock drains in any type of fill. Any nonacid- and nontoxic-forming rock type in this class is acceptable for use in federally-defined durable rock fills to comprise the 80% by unit volume durable rock component of the fill mass.</u> This rock has a measured strength of between 2 to 6 MPa, and a free-swell dilation at or below 4%.

Class III-Durable and strong rock drain material is suitable for **all** durable rock applications. It exhibits rock behavior during the soak-test (Ranking > 2), and has a rock strength measuring at or greater than 6 MPa, and has 4% swell or less;

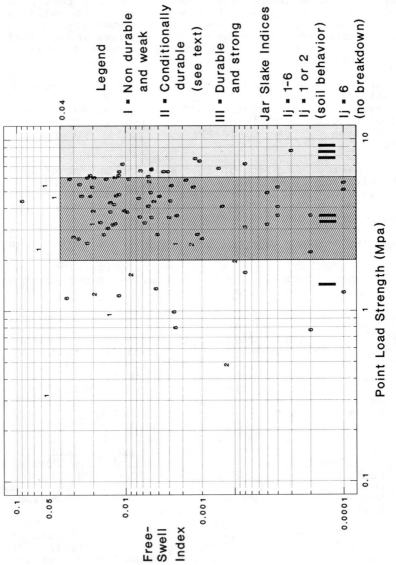

Figure 2. Strength-Durability Classification Of Jar Slaking

The range of 4 MPa depicted on figure 2 for the Class II rock corresponds to the typical decrease in rock compressive strengths when saturated samples are used instead of samples at the as-received moisture content. The OSM classification plot is employed to incorporate this observed loss of strength, and ensures that rock susceptible to such behavior (i.e. would be of low strength after water immersion) is not used as rock drain material.

The rating of rock swell employs the observed correlation in this study between soil behavior in the soak test and free-swelling values. Shale and mudstones tested in this study exhibited soil behavior when free-swell values exceeded 4% in all but one instance. This correlation indicates that seemingly low free-swelling magnitudes are responsible for not only volume increases along bedding planes, but actual rock disaggregation. Many samples disintegrated in both tests, preventing a measurement of free swelling.

Conclusion

The proposed OSM strength-durability classification system is a more accurate measure of rock strength and durability than present classification systems, particularly for rock materials of marginal durability, such as shale. Correlation with soak test results indicates that the swell and point load tests measure factors which are pertinent to the original concerns regarding rock durability in surface mining valley fills.

Acknowledgements

The authors appreciate the insight on testing procedures provided by Terry Stransky, Lane Tison, and Mark Buss of the U.S. Army Corps of Engineers, and James Smith of R&R International, Inc. Peer reviews of aspects of this research were performed by Donald Fohs, Stephen Forster, and Albert DiMillio of the Federal Highway Administration and Glenn Nicholson of the U.S. Army Corps of Engineers Waterways Experiment Station. However, the authors are solely responsible for the research findings and conclusions.

REFERENCES CITED

Andrews, D. E., Withiam, J. L., Perry, E. F., and Crouse, H.L., 1980. Environmental effects of slaking of surface mine spoils: eastern and central United States. Bureau of Mines, U.S. Department of the Interior, Denver, CO. Final Report, 247 pp.

Aufmuth, R.E., 1974. A systematic determination of engineering criteria for Bulletin of the Association of Engineering Geologists. Vol. 11, No. 3, pp. 235-245.

Aughenbaugh, N.B. and Bruzewski, R.F., 1976, Humidity effects of coal mine roof stability. BuMines Report No. OFR 5-78, Department of Interior, Washington, D.C., 160 p.

Bieniawski, Z. T., 1975. The point load test in geotechnical practice. Engineering Geology, Vol. 9, pp. 1-11.

Bjerrum, L., 1967, Progressive failure in slopes of overconsolidated plastic clay and clay shales. Journal of Soil Mechanics and Foundations Division, American Society of Civil Engineers, Vol. 93, No. SM5, pp. 1-19.

Broch, E. and Franklin, J.A., 1972. The point load strength test. International Journal of Rock Mechanics and Mining Sciences, Vol. 9, pp. 669-697.

Deere, D.U., and Miller, R.P., 1966. Engineering classification and index properties for intact rock, Technical Report No. AFWL-TR-65-116, Air Force Weapons Laboratory, Kirtland AFB, NM.

Duncan, N. 1969. Engineering geology and rock mechanics. 252 pp. Leonard Hill, London.

Duncan, N., Dunne, M.H., and Petty S., 1968. Swelling characteristics of rocks. Water Power. May 1968:185-192.

Franklin, J. A. 1970. Observations and tests for engineering description and mapping of rocks. Proceedings of 2nd International Congress on Rock Mechanics, Belgrade, Vol. 1, pp 1-3.

Grice, R.H., 1968. The effect of temperature-humidity on the disintegration of nonexpandable shales, Bulletin of the Association of Engineering Geologists, Vol. 5, pp. 69-77.

Hart, S.S., 1974, Potentially Swelling Soil and Rock in the Front Range Urban Corridor, Colorado, Environmental Geology No. 7, Colorado Geological Survey, Denver, CO, 23 p., 3 pls.

Hoek, E., and Bray, J., 1977, Rock slope engineering. Revised second edition. The Institution of Mining and Metallurgy, London, p. 227.

Holtz, W.G., 1959, Expansive clay - properties and problems. Colorado School of Mines Quarterly, Vol. 54, No. 4, pp. 90-125.

International Society for Rock Mechanics (ISRM), 1979. Commission on standardization of laboratory and field tests, suggested methods for determining water content, porosity, density, absorption and related properties and swelling, and slake-durability index properties. International Journal of Rock Mechanics and Mining Sciences & Geomechanics Abstracts, Vol. 16, No. 2, pp. 148-156.

Kirk, T.T., and Hall, G.E., 1982. Design considerations for valley fills. Green Lands, Vol. 12, No. 4, pp. 34-37.

Lama, R.D. and Vutukuri, V.S., 1978. Handbook on mechanical properties of rocks-testing techniques and results, Volume IV, Series on Rock and Soil Mechanics, Vol. 3, No. 3, pp. 274-316.

Lo, Y., Wai, R.S.C., Palmer, J.H.L., and Quigley, R.M., 1978. Time-dependent deformation of shaly rocks in southern Ontario. Canadian Geotechnical Journal, Vol. 15, pp. 537-547.

Lutton, Richard J. 1977. Design and construction of compacted shale embankments, Vol. 3, Slaking Indices for Design, FHWA-RD-77-1, 88 pp.

Moriwaki, Y., 1974. Causes of slaking in argillaceous materials. Ph.D. dissertation, University of California, Berkley, California, 291 pp.

Nascimento, U., Oliveira, R., and Graca, R., 1968, Rock swelling test. in Proceedings of the International Symposium of Rock Mechanics, Madrid.

Oakland, M.W. and Lovell, C.W. 1982. Classification and other standard tests of shale embankments: Joint Highway Research Project Report 82-4, Purdue University, West Lafayette, Indiana, 171 p.

Olivier, H.J., 1979. A new engineering-geological rock durability classification. Engineering Geology, Vol. 14, pp. 255-279.

Ollier, C.D., 1969. Weathering. Longman Group, Ltd., London, England, 304 p.

Quigley, R.M., Matich, M.A.J., Horvath, R.G., and Hawson, H.H., 1971, Swelling clay in two slope failures at Toronto, Canada. Canadian Geotechnical Journal, Vol. 8, pp 417-424.

Seedsman, R., 1986, The behaviour of clay shales in water. Canadian Geotechnical Journal, Vol. 23, pp. 18-22.

Strohm, W.E., Bragg, G.H., and Ziegler, T.W., 1978, Design and construction of compacted shale embankments, Vol. 5: Technical Guidelines, Report No. HWA-RE-78-141, Federal Highways Administration, Washington, D.C., 207 p.

Terzaghi, K. and Peck, B., 1967. Soil mechanics in engineering practice. Second Edition, New York. John Wiley & Sons, 729 p.

United States Code of Federal Regulations. 1989. Title 30, Mineral Resources, Chapter VII, Enforcement, Department of the Interior, Sections 817.71 and 817.73, pp. 320-324.

United States Federal Register. 1979. Surface Coal Mining and Reclamation Operations. Part II, Permanent Program Regulations, Vol. 44, No. 50, Tuesday March 13, 1979, pp. 15205-15206.

Vallejo, L.E., Welsh, R.A., Jr., and Robinson, M.K., 1989, Correlation between unconfined compressive and point load strengths for Appalachian Rocks, in Proceedings of the 30th U.S. Symposium on Rock Mechanics, Morgantown, West Virginia, June 19-22, 1989.

Van Eeckhout, E.M., 1976. The mechanisms of strength reduction due to moisture in coal mine shales. International Journal of Rock Mechanics and Mining Science and Geomechanics Abstracts. Vol. 13, pp. 61-67.

Venter, J.P., 1980. An investigation of the slake durability test for mudrocks used in road construction. Proceedings of the Seventh Regional Conference for Africa on Soil Mechanics and Foundation Engineering. Accra. pp. 201-206.

Determination and Presentation of Subsurface
Conditions for Design and Construction

Louis R. Frei[1]
Clarence O. Duster, Associate Member[2]

Abstract

A major problem in the design and construction of many
engineering projects is the lack of adequate geotechnical
information both for design and for determining conditions
to be encountered in required excavations. Although
excavation difficulties can represent major impacts to the
cost and construction schedules for a project, little
effort is usually made to define subsurface conditions in
terms useful for determining these difficulties.

Understanding geologic conditions and their significance
to the design and construction process is often a key to
the economics of a project. What is known, and more
importantly, what is not known about subsurface conditions
must be presented in terms that can be equated to both
design requirements and the economic consequences to
construction. During the design of a project, the quality
of the geotechnical data provided and the understanding of
the significance of that data to design will determine how
well the design is optimized for subsurface conditions.
The geotechnical data gathered must also address the
effects of the subsurface geologic conditions on the costs
of construction. Uncertainties that remain regarding
subsurface conditions and predicted response to
construction operations must be accommodated in the design
and communicated to prospective contractors so that the

[1]Consulting Engineering Geologist, (U.S. Bureau of
Reclamation, Retired), 24346 W. Currant Dr, Golden CO
80401

[2]Design Manager, U.S. Bureau of Reclamation, PO Box
25007, Denver, CO   80225

bidding process reflects the best understanding possible of subsurface conditions.   These concepts reflect the process currently implemented in the Bureau of Reclamation (Reclamation) (3).

Introduction

The selection of the topic of this symposium "Detection of and Construction at the Soil/Rock Interface" points out the need for more definitive information on the subsurface conditions that will be encoutered in construction. Often, designs are completed and contracts bid based on limited information on the subsurface conditions that will be encountered or how these conditions effect project costs.   The question posed by this symposium does not have a simple answer and should only be addressed as part of a total design and construction process.   This paper tries to define this overall design-construct process and show how high quality geotechnical data when obtained early is the process is useful in design and ulitmately provides, as far as possible, an answer to the question what and where is rock.

Design Exploration

Geotechnical data collection and analysis are particularly important in the design process.   The exploration phase is a dynamic process which requires continual refinement of data requirements, geologic interpretations, and design concepts as the information is obtained.   Personnel involved in the process must fully understand the need for and the significance of the data collected.   At the start of data collection, the geotechnical questions that must be answered should be outlined along with the reasons for the concerns, the proposed methods for resolving the questions, and the relevance of the data to the questions.

The primary objectives of geotechnical design data collection are:

• To gather the data necessary for sound engineering design.

• To present the data in a form amenable to analysis, design, and construction.

• To ensure that all appropriate data are available and used in design and throughout the construction process.

• To provide a sound basis for determining the conditions to be encountered during construction and anticipate the difficulty of accomplishing the work.

The first task in a geotechnical investigation program should be to evaluate all information on past geologic studies in the area. It is particularly important to obtain aerial photographs to evaluate the significance of geomorphic features to the intended design. The second task is to prepare a geologic map of the site using surface exposures and aerial photographs. This is the least expensive form of exploration and often the most valuable. After these tasks are accomplished, explorations should be directed toward defining geologic conditions brought out by the mapping to the fullest extent possible given time and cost constraints. An effective means for site specific evaluation of geologic data is to divide the site into geologic areas or units with similar engineering properties, define those engineering properties by additional investigation and testing, if appropriate, and present data in a form that can be easily understood and used in design.

The dynamic process of gathering data and continually performing three-dimensional interpretation and analysis of subsurface conditions ensures that the significance of the data both to the intended design and to further exploration needs is assessed. Data collection programs should be modified and design concepts revised to reflect the changing state of knowledge. Geologic drawings are the single most important product of the design data collection process since they are the primary tools used to develop and analyze subsurface conditions and relate these conditions to the design. Good geologic base maps and cross sections, developed at the same scale as the design drawings, are necessary to make the process efficient. These drawings, with modification and updating as significant new data are acquired, should be available throughout the design process for use by designers, geotechnical engineers, and engineering geologists.

When the design data collection process proceeds logically, interpretations are made as data are collected with a continual assessment of the data relative to design concepts and exploration requirements. The frequent analysis of geologic data, including interpretation and application to the proposed design, prevents unnecessary exploration and provides adequate high-quality design data at minimum cost. Collection of inadequate or inappropriate data can be prevented by modifying either the program or the design concept as the data are obtained and evaluated. This constant evaluation is the necessary key to an efficient and effective exploration program.

In addition to providing necessary information for analysis and design, explorations should also be oriented toward determining the ease or difficulty of accomplishing the required excavations or other construction activities. If both design and construction requirements are kept in mind, information can be obtained to satisfy all requirements and provide a basis for effectively estimating construction costs. For example, whenever construction equipment such as dozers or backhoes is used during explorations, the equipment performance should be recorded for each geologic unit encountered. Where practical, excavations should be taken deeper than may be necessary for design to explore the limit of excavation for a particular type of equipment. This can be accomplished for each defined geologic unit that will be encountered in excavation. Additional data on excavation characteristics or for other construction work can be obtained from the exploratory drilling. Drill penetration rates in different geologic units when combined and compared with other information can be useful in assessing construction difficulties. Also, permeability testing to determine conditions for design purposes should address potential dewatering problems.

Information Presented in Bid Documents

Design intent, critical geotechnical facts, and the design assumptions used in developing a design should be communicated through the bid documents to ensure successful construction of the project. In the past, Reclamation specifications were formulated to place all responsibility on prospective contractors to assess the significance of the design data gathered relative to the work to be accomplished. The following is an example of the disclaimers that were typically included in Reclamation specifications:

> The Government does not represent that the available cores, samples, logs, and other available geologic information show the conditions that will be encountered in performing the work, and the Government represents only that such information shows conditions encountered at the particular point from which such information was obtained. It is expressly understood that the making of deductions, interpretations, and conclusions from all the accessible factual information, including the nature of the materials to be excavated, the difficulties of making and maintaining the required excavations, and the difficulties of doing other work affected by the geology, and other subsurface conditions at the site of the work, are the contractor's sole responsibility.

These types of statements placed unrealistic requirements on bidders who were required not only to determine the conditions to be encountered but, for the most part, the intent of the design itself. Information regarding Reclamation's interpretation of the data and significance of these interpretations to the design generally was not made available to bidders.

Presently, Reclamation considers that since Reclamation determines the type, quality, and amount of design data necessary to produce adequate designs, they should also be responsible for the information presented in the specifications to define construction requirements. Since Reclamation routinely interprets the meaning and significance of the data in preparing the designs and specifications, and decides when it is more cost effective to move from data collection and design into construction, the specifications should recognize and reflect this process.

By obtaining good quality design data, effectively interpreting the data, and fully utilizing the information in the design, specifications can be prepared which accurately reflect the available knowledge and understanding of the site conditions and how these conditions may impact construction. The specifications can and should be used to convey to the bidders, the contractor, and construction management personnel not only the known conditions and work requirements but also those areas where uncertainties exist that may require design changes during construction.

This policy means that Reclamation provides in the specifications all geologic and the other design data considered useful to bidders and contractors. The data are interpreted to the fullest extent that is reasonable with relationships to the design demonstrated. The specifications explain what is known and what is assumed about geologic or other site conditions and how they are expected to affect construction, including the expected difficulty of accomplishing required excavations. Uncertainties with the data, or interpretation of the data, that may result in changes to the work requirements are clearly identified.

Full interpretation of the data provides bidders with the best possible definitions of the work requirements and the conditions that may be encountered in the work. With this information, bidders can prepare more realistic bids. The bidders should not have to account for risks associated with unknowns, since they can bid the work based on Reclamation's evaluation of anticipated conditions and identified uncertainties. With the anticipated and

potential impacts to the work clearly defined in the
specifications, a basis for determining and settling
changed conditions claims is established before
construction begins. The difference between the
anticipated conditions as defined in the specifications,
and the actual site conditions found during construction
can be defined with less difficulty and greater certainty.
If a contractor bids the work based on the anticipated
conditions, as presented in the specifications, the basis
for a claim should be the impact of any changed conditions
on the work, rather than a question of the existence of a
changed condition. This should greatly enhance and
expedite the process of resolving disputes regarding
changed site conditions.

The following excerpts from two recent Reclamation
specifications are provided to illustrate this concept.
The first example presents five statements from various
sections of a specifications (4) for the excavation of a
large cut to the final foundation grade for a powerplant
as shown in Figure 1:

> The rock at foundation grade is generally very
> hard, fresh to weathered, and moderately to
> intensely fractured. A large cliff upstream of
> the powerplant is parallel to joint set R and
> may continue in the subsurface into the
> powerplant yard area. Joints may form steep
> cliffs on the bedrock surface. Other joints
> are present in the granodiorite at the
> powerplant site which are not recorded in this
> survey. In particular, low-angle joints (less
> than 45° dip) not readily apparent on surface
> outcrops were intercepted by the drill holes.

> Permeabilities were not measured in the
> surficial deposits at the powerplant. However,
> permeability rates of 150 to 300 feet per day
> can be expected in the terrace and alluvial
> deposits. Permeability tests conducted in the
> Precambrian rock indicate secondary
> permeability (jointing and fracturing) is
> approximately 1.6 feet per day.

> (d) Excavation. - Design slopes of 1-1/2:1 in
> the surficial materials should be stable;
> however, some minor sloughing could occur
> during and following periods of heavy
> precipitation. Surficial materials can be
> excavated with a dozer unless large boulders
> are encountered. Igneous boulders up to 20
> feet in diameter are present in similar
> deposits along the Shoshone River.

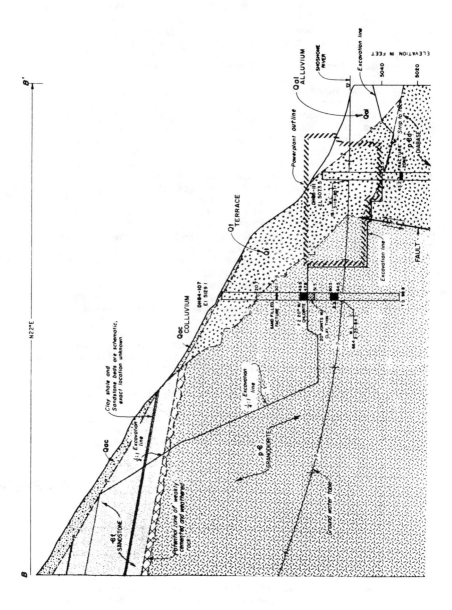

Figure 1.   Excavation for Powerplant - Section View

(e) Foundation. - The exact configuration of the top of the rock is not known in the powerplant area. The rock surface can be expected to deviate from that estimated on the geologic sections due to the limited data base used to predict the rock configuration. When rock is encountered it is expected to provide a suitable foundation; however, some weathered zones may have to be removed at selected locations. The buried scarp shown associated with the fault on the sections is based only on weathering characteristics upstream and may not exist in the powerplant area. Vertical cliffs may be present where the granitic rock has eroded along vertical joints. The foundation for the powerplant is predominantly very hard to extremely hard diabase with some granodiorite. The contact between these two materials will be less competent rock in a fault zone. The majority of the rock is slightly weathered to fresh in the powerplant area, minimizing the need for extensive cleanup. Some small areas of intensely weathered rock are expected especially near the fault.

The second example is from the specifications (2) for a modification to construct a spillway section at an existing embankment dam as shown in Figure 2:

Geotechnical and Construction Considerations

These specifications are the result of a spillway design change required because of a slide which occurred in the spring of 1989. The slide is located on the right side of the canyon immediately downstream of the dam, encompassing the entire original spillway alignment. The drawings show the location of the slide in relation to the proposed work. The slide represents the reactivation of a portion of a larger, old landslide and was caused by excavating the original stilling basin at the toe of the slide and stockpiling material higher on the slide mass.

In order to stabilize the slide, the original stilling basin excavation was backfilled and a stability berm constructed on the toe of the slide which included a ramp for the new spillway. Drain holes were drilled in the slide mass to relieve uplift pressures under

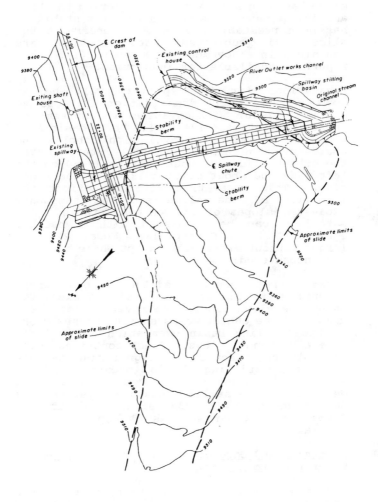

Figure 2.   New Spillway Design After Slide Stabilization

the slide. Instrumentation has shown that the
slide has been stable since completing the berm
and drains, and it is considered that stability
has been reestablished.Details concerning the
slide and subsequent stabilization are
contained in a November 1989 report (1). This
report is available for inspection.

Excavations for the spillway will be in
compacted earthfill from station 10+34.53 to
station 17+40. Excavation downstream of
approximate station 17+40 will be in basin fill
deposits and stream gravel with cobbles and
boulders. The excavation will intersect the
toe of the slide at about station 17+60 on the
left side and about station 18+40 on the right
side. This excavation will not effect the
stability of the slide and the temporary
excavated slopes of 1:1 will be stable. All
materials can be excavated with a backhoe
equivalent to a Caterpillar, Model 225, or
larger without difficulty except where boulders
are encountered.

Water will be encountered in the old stream
channel where the stilling basin is excavated
below the elevation of the river outlet works
channel. The outlet works channel passes close
to the stilling basin excavation on the left
side and channel flows, if not contained, will
probably be a source of significant seepage
into the stilling basin excavation. The
stability of the alluvium and basin fill
deposits in the stilling basin excavation will
depend on how well the area is dewatered.
Seepage from other sources is expected to be
minimal.

Diversion and Care of the Stream and Removal of
Water From Foundations

The items for diversion and care of the stream
and removal of water from foundations are
interrelated and must be considered together.
The stilling basin excavation encroaches into
the outlet works channel as shown on the
drawings, therefore the method of containing
the stream in this area will influence
dewatering of the excavation. Both the
quantity of water to be removed and the method
of removal will be influenced by the nature of
the streambed materials, how the stream is

contained or diverted, the length of any containment, and river outlet works releases.

It has been experienced that groundwater levels in the channel area reflect the tailwater of the stream. Inclinometer hole I-4 best reflects this condition. Water levels in I-4 have remained relatively constant since the inclinometer was installed in July 1989. The elevation of the reservoir should have a minimal affect on the water levels encountered in the river channel area in the vicinity of stilling basin excavation. Available data also indicates that locally, the groundwater is flowing towards the original stream channel. The toe drain outfall for the dam located just left of the river outlet works outlet structure will also contribute to the water requiring diversion.

The depth and lateral extent of alluvium in the original channel is unknown. The original channel enters the left side of the stilling basin excavation at about station 17+85 and exits the stilling basin excavation at about spillway centerline. During the previous construction, excavation for the outlet works control structure exposed shallow (about 1- to 4-feet deep) alluvium underlain by basin fill material consisting of clayey gravels, tuffs and some cemented materials. The alluvium within the original channel is a heterogeneous mixture of clay, silt, sand, gravel, cobbles, and boulders. The alluvium in the original steam channel is assumed to be highly permeable and the basin fill material should be assumed to have low to moderate permeability. The alluvium within the original creek channel could be cut off to reduce seepage into the spillway excavation. The volume of water to be pumped is unknown. Trenching in the stilling basin area combined with pump testing should provide valuable information for designing an effective dewatering system.

During the previous construction, the excavation for the outlet works control structure, located upstream from the spillway stilling basin excavation, was successfully dewatered; using shallow wells, pumps, and french drains.

Consistent with the policy of presenting interpretations of site conditions that will affect construction,

Reclamation is also eliminating the terms "rock" and "common" excavation to define excavation difficulties. The current procedure of providing the bidder with estimates of the difficulty of accomplishing required excavations recognizes that conditions are usually unique to each job and the difficulty of excavation can be expressed in different ways in specifications. The method of expression depends on the number and variety of materials to be excavated, and the type of excavation required. In some cases, difficulty of excavation may be simply quantified as a depth from the ground surface or a percentage of the total excavation. In more complex cases, difficulty of excavation can be related to different geologic units which have complex spacial relationships to the excavation. The definition of difficulty or type of excavation is defined as it specifically relates to each specifications, and expressed in terms most appropriate to those requirements.

The designers'/geologists' assessment of their ability to adequately describe the difficulty of excavation or suitability of earth materials for use in engineering structures is also expressed. This assessment is based on the quality and completeness of the design data and the complexity of the geology. This information provides the bidders, the contractor, and the construction management staff an understanding of the degree of certainty that can be placed on the information presented in the specifications.

Design During Construction

The construction process can also be improved through more effective interaction between design and construction staffs. The construction process should be a continuation of designs, although at a more rapid pace. At the start of construction, designs and specifications reflect the best interpretation of the geologic conditions. Construction affords the opportunity to obtain the data necessary to either verify the geologic conditions assumed during design or modify the designs if conditions are not as anticipated. During construction, these data are obtained mainly from geologic mapping of required excavations.

In order to achieve an effective process of design verification or modification and avoid delay in construction, geologic mapping must be accomplished as rapidly as possible. Information must be integrated with the preconstruction geologic design data in a form amenable to analysis and verification of the design. To facilitate rapid and effective evaluations, the information should be presented at the same scale and in

the same format as that used in design. Progressive
revision of the drawings to reflect as-excavated
conditions will, in many cases, allow changes in geologic
conditions to be predicted and the design modified ahead
of the actual excavation. Even if changes in geologic
conditions are not predicted prior to actual excavation,
geologic mapping concurrent with the excavations will
provide the data to rapidly evaluate the significance and
implication of changes and the effect on construction.

For this process to be successful, a mechanism for rapid
analysis of differing geologic conditions and
implementation of design changes should be developed.
Several requirements can be presented in the
specifications to facilitate this process during
construction:

• Excavated surfaces should be cleaned immediately
after exposure to allow for geologic mapping. Additional
cleanups should be provided for in case unanticipated
situations arise.

• Topography of excavated surfaces should be
obtained immediately after excavation to allow for rapid
geologic mapping and a progressive comparison between the
design and the actual excavated surface.

• Excavations should be cleaned an appropriate
distance beyond areas requiring foundation acceptance to
allow sufficient time for analysis and evaluation of
actual conditions.

• Exploration during construction should be provided
for on an as-needed basis.

Reclamation has adopted a policy of foundation acceptance
to ensure that the intent of the design is accomplished
during construction. The policy applies to all critical
structures, and, in particular dam foundations. The
foundation acceptance procedures involve examination and
approval of the final foundation by appropriate design
team and construction representatives, prior to fill or
concrete placement, to ensure conformance with the design
intent. Each foundation acceptance is documented by a
memorandum which is prepared at the construction site and
concurred with by the project construction engineer.

Conclusion

The procedures presented, if implemented, can provide a
superior product at the lowest possible cost. In most
cases it is poor data or ineffective use or understanding
of the geotechnical data by the designers and/or

contractor that results in an incomplete understanding or misunderstanding of the site conditions and their effect on design and construction. Obtaining high quality data, interpreting the significance of the data for both design and construction, and presenting those interpretations in the specifications documents can provide not only a high quality product but a means of minimizing disputes regarding changed site conditions. It also provides a basis for defining the magnitude and impact of changes when they occur.

APPENDIX. - REFERENCES

[1] Frei, L.R., and Von Thun, J.L., "Costilla Dam, Costilla Slide Stabilization," *Report to New Mexico Interstate Stream Commission*, Nov., 1989.

[2] New Mexico Interstate Stream Commission, "Costilla Dam Modification Completion," *Document No. 1-553.01-2*, New Mexico, Apr., 1991.

[3] Smart, J.D., Frei, L.R., and J.L. Von Thun, "Geotechnical Engineering in the Design and Construction Process," *The Art and Science of Geotechnical Engineering, at the Dawn of the Twenty - First Century, A Volume Honoring Ralph B. Peck*, Prentice Hall, Englewood Cliffs, New Jersey, 1989, pp. 336-348.

[4] US Bureau of Reclamation, "Buffalo Bill Powerplant," *Solicitation/Specification 6-SI-60-02070/DC-7682*, Wyoming, Jul., 1986.

SUBJECT INDEX
Page number refers to first page of paper.

AUTHOR INDEX
Page number refers to first page of paper.

Date Due

BRODART, CO.   Cat. No. 23-233-003   Printed in U.S.A.